Coastal Engineering
Theory and Practice

ADVANCED SERIES ON OCEAN ENGINEERING

Editor-in-Chief
Philip L-F Liu (*Cornell University and National University of Singapore*)

Published

Vol. 47 Coastal Engineering: Theory and Practice
by Vallam Sundar and Sannasi A. Sannasiraj

Vol. 46 Tsunami: To Survive from Tsunami (Second Edition)
by Tomotsuka Takayama, Kazumasa Katoh, Fumihiko Imamura, Yoshiaki Kawata, Susumu Murata and Shigeo Takahashi

Vol. 45 Ocean Surface Waves: Their Physics and Prediction (Third Edition)
by Stanisław Ryszard Massel

Vol. 44 Satellite SAR Detection of Sub-Mesoscale Ocean Dynamic Processes
by Quanan Zheng

Vol. 43 Japan's Beach Erosion: Reality and Future Measures (Second Edition)
by Takaaki Uda

Vol. 42 Theory and Applications of Ocean Surface Waves (Third Edition)
by Chiang C. Mei, Michael Aharon Stiassnie and Dick K.-P. Yue

Vol. 41 Dynamics of Coastal Systems (Second Edition)
by Job Dronkers

Vol. 40 Design and Construction of Berm Breakwaters
by Jentsje Van der Meer and Sigurdur Sigurdarson

Vol. 39 Liquefaction Around Marine Structures
by Mutiu Sumer

Vol. 38 An Introduction to Hydraulics of Fine Sediment Transport
by Ashish J. Mehta

Vol. 37 Computational Wave Dynamics
by Hitoshi Gotoh, Akio Okayasu and Yasunori Watanabe

Vol. 36 Ocean Surface Waves: Their Physics and Prediction (Second Edition)
by Stanislaw R. Massel

Vol. 35 Dynamics of Water Waves: Selected Papers of Michael Longuet-Higgins
edited by S. G. Sajjadi

Vol. 34 Coastal Dynamics
by Willem T. Bakker

Vol. 33 Random Seas and Design of Maritime Structures (Third Edition)
by Yoshimi Goda

*For the complete list of titles in this series, go to https://www.worldscientific.com/series/asoe

Advanced Series on Ocean Engineering — Volume 47

Coastal Engineering
Theory and Practice

Vallam Sundar
S. A. Sannasiraj

IIT Madras, India

NEW JERSEY · LONDON · SINGAPORE · BEIJING · SHANGHAI · HONG KONG · TAIPEI · CHENNAI · TOKYO

Published by

World Scientific Publishing Co. Pte. Ltd.
5 Toh Tuck Link, Singapore 596224
USA office: 27 Warren Street, Suite 401-402, Hackensack, NJ 07601
UK office: 57 Shelton Street, Covent Garden, London WC2H 9HE

British Library Cataloguing-in-Publication Data
A catalogue record for this book is available from the British Library.

Advanced Series on Ocean Engineering — Vol. 47
COASTAL ENGINEERING
Theory and Practice

Copyright © 2019 by World Scientific Publishing Co. Pte. Ltd.

All rights reserved. This book, or parts thereof, may not be reproduced in any form or by any means, electronic or mechanical, including photocopying, recording or any information storage and retrieval system now known or to be invented, without written permission from the publisher.

For photocopying of material in this volume, please pay a copying fee through the Copyright Clearance Center, Inc., 222 Rosewood Drive, Danvers, MA 01923, USA. In this case permission to photocopy is not required from the publisher.

Cover illustration: Coastal protection along the Chennai coast

ISBN 978-981-3275-90-4

For any available supplementary material, please visit
https://www.worldscientific.com/worldscibooks/10.1142/11148#t=suppl

Printed in Singapore

Preface

The ocean occupies about 70% of the earth's surface which is bordered by the irregular boundaries of land areas called coastline. These coastal stretches are usually densely populated in comparison to the landlocked regions, primarily for trade and commerce through important/predominant sea routes and hence important cities are mostly formed along the coasts. Owing to the large scale expansions, several infrastructural developments have been witnessed along the shoreline, thereby a need to protect the life and property of the coastal community arises. It is necessary to understand the dynamics and physics of the coasts in general to propose protection measures. It is also noteworthy to mention that the coastal regions are more prone to natural hazards and hence specific measures need to be taken in order to prevent hazards. The primary objective of this book is to serve the senior undergraduate and graduate students, and researchers engaged in the field of ocean/coastal engineering to understand the physics/dynamics of coasts in response to the wind wave action with a number of worked out examples and cases. This book is self-contained, since all the representations have been clearly defined in the text. Further, the sequential arrangement of chapters discussed in the book is well planned and drafted such that the understanding of each of the chapters can be greatly enhanced by reading its preceding chapter. It encompasses the fundamentals of the subject with sufficient description and illustrations.

It is also expected to be of great help to professionals working in major ports, harbours, coastal engineering, consultancy agencies, etc.

The *Introduction* facilitates one to understand the different terminologies pertaining to the study of coastal engineering and gives an insight into the different types of wave deformation that could potentially alter the coastal geo-morphology. A brief introduction about the Indian coast with special focus on the coastal states of Tamil Nadu and Kerala are discussed in this chapter.

In order to grasp the concept of sediment motion in response to the combined action of wind, wave, current, tide, etc., one should possess a strong foundation of basics in sediment characteristics. To facilitate this, the chapter on **Characteristics and motion of sediments** gives a brief elucidation about the basic definitions of soil mechanics which are crucial to understand the behaviour of sediments in the coastal/marine environment. Sediment motion is the most significant aspect of coastal engineering, which is introduced to the readers in this chapter.

A sizeable share of this book lays emphasis on **Sediment transport**, which is the central interest and major focus. A notable account on *Littoral transport* followed by profound explanation and inferences of dynamic nature of the cross-shore and longshore sediment transport is included in this section. The different mathematical and empirical relations derived to quantify the volume and direction of sediment transport across a given coastal stretch are also discussed. A detailed summary of the various methods proposed for sediment transport estimation over the years is discussed. The novel concept of sediment cell and the changes in pattern of sediment transport in the wake of extreme events along with case studies and worked examples are reviewed.

The chapter on **Coastal erosion and protection measures with case studies** discusses in detail the physics of sediment erosion and the measures to combat erosion problems. Apart from the broad classification of hard and soft measures for coastal protection, a detailed examination of the numerous well-established and a few unconventional measures are detailed with real time case studies. A comparison of the performance assessment of the hard and soft measures is made to facilitate understanding of the implication of employing the same in the field. Additionally, the significance and functioning of tidal inlets are discussed.

A detailed account on **Rubble mound structures** with special focus on breakwaters in this chapter includes its criteria of selection and design principles. It vastly aids a pursuing student/learner to streamline one's understanding of the various component layers/units of a breakwater cross section and its evolution to adapt as a modern-day structure in contrast to its conventional norms. The development of various concrete armour layer units over the years to address the various failure mechanisms and ease of transportation and placing are dealt with in this chapter, in addition to discussing the damage assessment on the armour units.

The chapter on **Wave run-up and overtopping** introduces their effect to the readers and the importance of the same in the design of coastal structures. These on-shore phenomena dictate the crest elevation and the seaward slope of coastal structures.

The phenomenon of scour is an important failure mechanism of structures in coastal and marine environment, the causes and effects of which are discussed in the chapter on *Scour around marine structures*. The various parameters that lead to scour and different types of scour formation are discussed herewith to notify the engineers regarding the importance in design against scour failure. In addition, a brief introduction to pipelines and scour related to pipelines along side scour protection and bed shear stresses are detailed.

The understanding of various concepts in coastal engineering is essential to dictate appropriate design principles. The chapter on *Design of coastal structures* discusses the various methods of computing wave forces induced on structures due to breaking and non-breaking waves. The design principles and formulation of various layers of rubble mound structures are detailed with examples. In addition the design principles of retaining structures, sheet piles, marine piles, etc. are briefed.

This book would fulfill its wholesome objectives and become complete only if the concept of *physical* and *numerical modeling* are discussed. Thus at the last leg, the various methods and techniques for assessing the performance of coastal structures are briefly presented. Any proposed protection measure can be potentially evaluated for their functioning at calm or extreme events and possible failure mechanisms using the techniques discussed through laboratory measurements.

The references included in this book are vast and extensive, yet they would furnish more exhaustive topics pertaining to this field of study.

Contents

Preface v

1. Introduction 1

1.1 Background 1
1.2 Behaviour of Waves 2
    1.2.1 Shoaling 3
    1.2.2 Refraction 5
    1.2.3 Diffraction 6
    1.2.4 Breaking 7
    1.2.5 Reflection 8
1.3 The Indian Coast 9
    1.3.1 General 9
    1.3.2 Tamil Nadu 12
    1.3.3 Kerala 15
1.4 Summary 18
References 18

2. Characteristics and Motion of Sediments 20

2.1 Introduction 20
2.2 Sediment Classification 21
2.3 Particle Size 21
    2.3.1 Soil classification 21
2.4 Plasticity 25
2.5 Shape 26
2.6 Fall Velocity (v_f) 26
2.7 Angle of Repose 28
2.8 Effect of Temperature 28
2.9 Effect of Sediment Concentration 28
2.10 Effect of Turbulence 29
2.11 Permeability and Porosity 29

	2.12	Bulk Creep	30
	2.13	Sediment Motion	30
		2.13.1 General	30
		2.13.2 Incipient sediment motion	31
	2.14	Liquefaction of Sands	32
		2.14.1 General	32
		2.14.2 Soil types susceptible to liquefaction	33
	2.15	Shields Curve	33
	References		37
3.	**Sediment Transport**		**38**
	3.1	Introduction	38
	3.2	Modes of Sediment Transport	38
		3.2.1 General	38
		3.2.2 Description of the threshold of movement	39
		3.2.3 Load transport general approach	40
	3.3	Littoral Transport	40
		3.3.1 General	40
	3.4	Definitions	41
	3.5	Driving Forces	43
	3.6	Radiation Stresses	43
	3.7	Cross-shore Sediment Transport	50
	3.8	Longshore Sediment Transport	53
		3.8.1 Nearshore currents responsible for sediment transport	53
		3.8.2 Phenomena of littoral drift	54
		3.8.3 Estimating longshore transport	56
		3.8.4 CERC method	56
		3.8.5 Calculation of P_{ls} using LEO data	60
		3.8.6 Method of Kamphuis [1991]	61
		3.8.7 Sediment distribution across the surf zone	62
		3.8.8 Van Rijn method	64
	3.9	Sediment Cell Concept and Its Application	64
	3.10	Sediment Transport During Extreme Events	67
		3.10.1 General	67
		3.10.2 Case study	69
		3.10.3 Effect of sedimentation during extreme events	74
	References		95

4. Coastal Erosion and Protection Measures Including Case Studies — 97

- 4.1 Introduction — 97
- 4.2 Erosion Process — 97
- 4.3 Causes for Coastal Erosion — 98
- 4.4 Strategy for Coastal Protection — 99
- 4.5 Coastal Protection Measures — 102
  - 4.5.1 General — 102
  - 4.5.2 Hard measures — 102
    - 4.5.2.1 Hard structures — 103
    - 4.5.2.2 Soft structures — 111
  - 4.5.3 Soft measures — 114
    - 4.5.3.1 Beach nourishment — 114
    - 4.5.3.2 Placement of sand and borrow site — 116
    - 4.5.3.3 Methods — 116
    - 4.5.3.4 Vegetation cover — 118
- 4.6 Case Studies — 119
  - 4.6.1 Concept generation — 119
  - 4.6.2 Tamil Nadu (groin field) — 121
    - 4.6.2.1 North of Chennai Harbour (transitional groin field) — 121
    - 4.6.2.2 Vaan Island (submerged artificial reefs) — 124
  - 4.6.3 Kerala coast — 128
    - 4.6.3.1 General — 128
    - 4.6.3.2 Behaviour of seawalls prior to and after 2008 — 130
    - 4.6.3.3 Failure of seawalls — 132
    - 4.6.3.4 Behaviour of groin fields — 134
    - 4.6.3.5 Artificial beach nourishment — 137
    - 4.6.3.6 Geosynthetic products as coastal protection measure — 139
- 4.7 Assessment of Hard and Soft Measures — 142
- 4.8 Tidal Inlets — 143
  - 4.8.1 General — 143
  - 4.8.2 Tidal flushing — 144
  - 4.8.3 Stability of an inlet — 145
  - 4.8.4 Stabilisation of tidal inlets — 147

	4.8.5 Crater — Sink sand transfer system	152
4.9	Case Studies on Tidal Inlets	153
References		158

5. Rubble Mound Structures — 159

5.1	Introduction	159
5.2	Types of Breakwaters	160
5.3	Criteria for Breakwater Selection	163
5.4	Design Principles of Rubble Mound Structures	169
5.5	Concrete Armour Layer Units	170
	5.5.1 General	170
	5.5.2 Randomly placed armour units — stability factors weight and interlocking	170
5.6	Kolos	172
	5.6.1 General	172
	5.6.2 Dolos vs. Kolos	175
	5.6.3 Finite element model	175
	5.6.4 Results of analysis	179
5.7	Stability of CAUs	180
	5.7.1 Damage assessment	182
	5.7.2 Number of units method	183
References		184

6. Wave Run-up and Overtopping — 185

6.1	Introduction	185
6.2	Wave Run-up	186
	6.2.1 General	186
	6.2.2 Recent run-up equation	191
6.3	Wave Overtopping	194
	6.3.1 General	194
	6.3.2 Calculation of overtopping rates	195
	6.3.3 Complex slopes	202
	6.3.4 Designing for overtopping	204
6.4	Summary	206
References		208

7. Scour Around Marine Structures — 210

7.1	Introduction	210
7.2	Mechanism of Scour	211

		7.2.1	General	211
		7.2.2	Fluid mechanism of scour	212
		7.2.3	Scour due to steady current	213
		7.2.4	Scour due to waves	214
		7.2.5	Scour due to simultaneous action of waves and current	214
	7.3	Sediment Dynamics of Scour		215
	7.4	Types of Scour		215
		7.4.1	General	215
		7.4.2	General scour	216
		7.4.3	Local scour	216
		7.4.4	Degradation scour	217
		7.4.5	Boat scour	218
		7.4.6	High-head scour	218
		7.4.7	Global or dishpan scour	218
	7.5	Scour Failures and Evolution		219
	7.6	Scour Due to Vertical Walls		220
	7.7	Pipelines		222
		7.7.1	General	222
		7.7.2	Scour around pipelines	222
		7.7.3	Scour around pipelines due to current action	224
		7.7.4	Scour due to waves	226
		7.7.5	Scour due to waves and currents	226
	7.8	Maximum Scour Depth		227
	7.9	Scour Protection		228
		7.9.1	General	228
		7.9.2	Rip rap rock fill	229
		7.9.3	Protective mattress	229
		7.9.4	Buried toe	229
		7.9.5	Sand bags or grout filled bags	230
		7.9.6	Concrete grout	230
		7.9.7	Structural improvements	230
	7.10	Bed Shear Stress		230
		7.10.1	Bed shear stress due to waves	230
		7.10.2	Current related bed shear stress	231
		7.10.3	Combined wave and current shear stress	232
	References			233

8. Design of Coastal Structures — 235

- 8.1 Introduction — 235
- 8.2 Non-breaking Wave Forces — 236
- 8.3 Wave Forces on Walls and Rubble Mound Structures — 241
  - 8.3.1 Rubble mound structures — 241
  - 8.3.2 Pressure distribution on an overtopped wall — 242
  - 8.3.3 Minikin's method for a wall on a low rubble mound — 243
  - 8.3.4 Wall of rubble foundation — 245
  - 8.3.5 Breaking wave forces on vertical walls — 245
  - 8.3.6 Wall on a rubble mound — 247
  - 8.3.7 Wall of low height — 247
- 8.4 Goda's Method of Breaking Wave Force (1974) — 251
- 8.5 Rubble Mound Structures — 252
  - 8.5.1 General — 252
  - 8.5.2 Armour layer — 253
  - 8.5.3 Underlayer — 254
  - 8.5.4 Core layer — 254
  - 8.5.5 Toe mound — 254
  - 8.5.6 Thickness of armour and underlayer — 255
  - 8.5.7 Crest elevation — 255
  - 8.5.8 Filter layer — 255
- 8.6 Vertical and Composite Structures — 255
- 8.7 Retaining Structures — 259
  - 8.7.1 Gravity retaining walls — 259
  - 8.7.2 Sheet pile walls — 260
  - 8.7.3 Anchored earth structures — 263
- 8.8 Marine Piled Structures — 264
- 8.9 Problems — 266
- References — 293

9. Physical Modeling — 295

- 9.1 Introduction — 295
- 9.2 Dimensional Analysis — 295
  - 9.2.1 General — 295
  - 9.2.2 Rayleigh's method — 298
  - 9.2.3 Buckingham's pi theorem — 299

9.3	Model Analysis	301
	9.3.1 General	301
	9.3.2 Complete similarity	302
	9.3.3 Applications of model analysis	302
9.4	Principles of Similitude	303
	9.4.1 General	303
	9.4.2 Similitude in hydrodynamic problems	303
	9.4.3 Types of similitude	303
9.5	Scale Effects	305
9.6	Model Laws	305
9.7	Case Studies	307
References		314

10. Numerical Modelling — 315

10.1	Introduction	315
10.2	Need for Numerical Models	315
10.3	Mathematical Description of Flows: Governing Equations	317
	10.3.1 Continuity equation	317
	10.3.2 Momentum equation	317
10.4	Discretization	318
	10.4.1 Discretization techniques	319
	10.4.1.1 Finite difference method (FDM)	319
	10.4.1.2 Finite volume method (FVM)	319
	10.4.1.3 Finite element method (FEM)	319
10.5	Numerical Wave Modelling	320
	10.5.1 Wave spectral models	320
	10.5.2 Test case: Wave propagation over constant depth bathymetry	321
	10.5.3 Test case: Wave prediction over Bay of Bengal during a cyclone	323
10.6	Mild-Slope Equation (MSE) Wave Models	327
	10.6.1 Solution of MSE	329
10.7	Boussinesq Approximation	333
	10.7.1 Boussinesq equations	334
	10.7.2 Shallow-water equation wave models	336
References		336

Index   339

Chapter 1

Introduction

1.1 Background

Coastal zone is an area of interaction between land and sea which include renewable and non-renewable resources. Hence, the interaction between various natural processes and human activities are important factors in the coastal area for its sustainable development. The dynamic coastal environment is dictated by the major driving forces due to wind, waves, tide and current and the resilience characteristics of the coastal morphology. The need for the engineering development necessitates the development of many systems such as coastal protection measures, harbour structures, river training walls, intake well and outfall systems which may further influence the coastal behaviour. Among all the parameters needed for the design of structures in the marine environment, the effect of ocean waves play a dominant role in dictating the system design.

Beaches are formed by the sediments driven from the seabed by the ocean waves that undergo different types of transformation prior to breaking near the coast. The sediments in the seabed are picked up by the waves when they feel the sea bottom. Its movement then depends on the characteristics of the sediments as well as that of the waves. Sandy beaches extend from the outermost breakers to the landward limit of normal wave and swash action. A typical beach profile and the terminology adopted in defining are shown in Fig. 1.1. The wave energy incident on the beach dictates its profile. The beach in general constitutes the backshore, the foreshore, the portion sloping toward the ocean; and the nearshore, There could be a line of breakers or a region of wave breaking, the region, where, the wave induced sediments are in suspension that are driven along the shore by alongshore currents or tend moving along the cross shore. The movement of sediments along the coast will be dominated in the surf zone, that is, between the shoreline and the breaker zone.

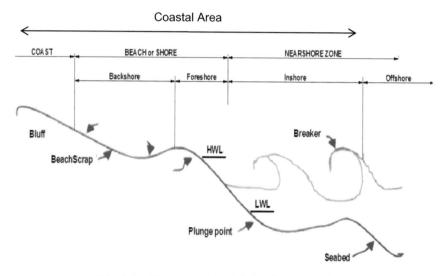

Fig. 1.1   Terms used in defining beach profiles.

In general, high waves with short periods cause the beach to erode, and the berm sand is shifted offshore to the bar. Low waves with longer periods, move sand from the bar and return it to the berm.

1.2   Behaviour of Waves

Ocean waves are oscillatory in motion with a height, H and length, L propagating with a speed, C in a water depth, d. The time taken for a wave to travel one wavelength defined as wave period is "T" the pictorial representation of which is as shown in Fig. 1.2. Linear wave theory is the basic theory of ocean surface waves used in ocean and coastal engineering and naval architecture. The *sine* (or *cosine*) function defines what is called a regular wave.

A summary of the linear wave theory as per the Coastal Engineering manual (2002) is reproduced in Table 1.1.

Ocean waves traveling in the offshore with no net gain or loss of energy will propagate in straight lines with a constant speed. However, while they propagate from offshore towards the coast undergo deformation due to the variations in the bathymetry, nature of the seabed like friction and presence of structures, interaction of currents and winds. These deformations in general are broadly classified as shoaling, refraction, diffraction, reflection and breaking which are briefly discussed herein. For a comprehensive

1. Introduction

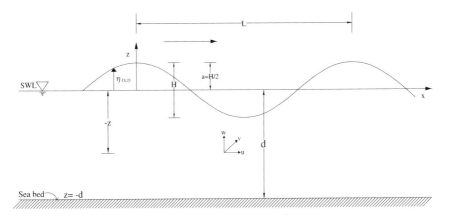

Fig. 1.2 Definition sketch for wave motion.

discussion on the basics of wave motion the readers are suggested to refer Sundar (2015) and any other books.

1.2.1 Shoaling

When waves propagate over a sloping seabed with bottom depth contours being parallel, its length decreases. The power transmission in a wave is proportional to $C_g H^2$. In case of pure shoaling, it is considered that the energy transfer is in the direction of propagation and there is no transfer of power in the lateral direction, i.e., normal to the wave direction. On assumption that there is no energy loss and by equating the average energy flux in deep-water, $\overline{P}_0 = 1/2 E_0 C_0$ to that in shallow waters, $\overline{P} = \overline{E} C_g$ the relationship between the wave heights in deep and shallower waters can be defined as

$$\frac{H}{H_0} = \sqrt{\frac{C_0}{C} \cdot \frac{1}{2n}} = K_s \tag{1.1}$$

where $n = \frac{1}{2}[1 + \frac{2kd}{\sinh 2kd}]$.

The relationship in Eq. (1.1) shows that as the waves propagate from deep to shallow waters, its height will gradually increase. The variations of K_s, n, C/C_0 and d/L as a function of d/L_0 for small amplitude waves are shown in Fig. 1.3. Herein, L, $L_0 = 1.56T^2$, $C = L/T$ and d, are the wavelength in a given depth, deep water wave length, celerity or speed of the wave and water depth respectively. T is the wave period.

Table 1.1. Basics of linear wave theory.

RELATIVE DEPTH WAVE PARAMETER	SHALLOW WATER $\frac{d}{L} < \frac{1}{20}$	TRANSITIONAL WATER $\frac{1}{20} < \frac{d}{L} < \frac{1}{2}$	DEEP WATER $\frac{d}{L} > \frac{1}{2}$
1. Wave profile	Same as $\rightarrow$	$\eta = \frac{H}{2}\cos\left[\frac{2\pi x}{L} - \frac{2\pi t}{L}\right] = \frac{H}{2}\cos\theta$	$\leftarrow$ Same as
2. Wave celerity	$C = \frac{L}{T} = \sqrt{gd}$	$C = \frac{L}{T} = \frac{gT}{2\pi}\tanh\left(\frac{2\pi d}{L}\right)$	$C = C_0 = \frac{L}{T} = \frac{gT}{2\pi}$
3. Wavelength	$L = T\sqrt{gd} = CT$	$L = \frac{gT^2}{2\pi}\tanh\left(\frac{2\pi d}{L}\right)$	$L = L_0 = \frac{gT^2}{2\pi} = C_0 T$
4. Group velocity	$C_g = C = \sqrt{gd}$	$C_g = nC = \frac{1}{2}\left[1 + \frac{4\pi d/L}{\sinh(4\pi d/L)}\right]C$	$C_g = \frac{1}{2}C = \frac{gT}{4\pi}$
5. Water particle velocity a) Horizontal b) Vertical	$u = \frac{H}{2}\sqrt{\frac{g}{d}}\cos\theta$ $w = \frac{H\pi}{T}\left(1 + \frac{z}{d}\right)\sin\theta$	$u = \frac{H}{2}\frac{gT}{L}\frac{\cosh[2\pi(z+d)/L]}{\cosh(2\pi d/L)}\cos\theta$ $w = \frac{H}{2}\frac{gT}{L}\frac{\sinh[2\pi(z+d)/L]}{\cosh(2\pi d/L)}\sin\theta$	$u = \frac{\pi H}{T}e^{\frac{2\pi z}{L}}\cos\theta$ $w = \frac{\pi H}{T}e^{\frac{2\pi z}{L}}\sin\theta$
6. Water particle acceleration a) Horizontal b) Vertical	$a_x = \frac{H\pi}{T}\sqrt{\frac{g}{d}}\sin\theta$ $a_z = -2H\left(\frac{\pi}{T}\right)^2\left(1 + \frac{z}{d}\right)\cos\theta$	$a_x = \frac{g\pi H}{L}\frac{\cosh[2\pi(z+d)/L]}{\cosh(2\pi d/L)}\sin\theta$ $a_z = -\frac{g\pi H}{L}\frac{\sinh[2\pi(z+d)/L]}{\cosh(2\pi d/L)}\cos\theta$	$a_x = 2H\left(\frac{\pi}{T}\right)^2 e^{\frac{2\pi z}{L}}\sin\theta$ $a_z = -2H\left(\frac{\pi}{T}\right)^2 e^{\frac{2\pi z}{L}}\cos\theta$
7. Water particle displacement a) Horizontal b) Vertical	$\xi = -\frac{HT}{4\pi}\sqrt{\frac{g}{d}}\sin\theta$ $\zeta = \frac{H}{2}\left(1 + \frac{z}{d}\right)\cos\theta$	$\xi = -\frac{H}{2}\frac{\cosh[2\pi(z+d)/L]}{\sinh(2\pi d/L)}\sin\theta$ $\zeta = \frac{H}{2}\frac{\sinh[2\pi(z+d)/L]}{\sinh(2\pi d/L)}\cos\theta$	$\xi = -\frac{H}{2}e^{\frac{2\pi z}{L}}\sin\theta$ $\zeta = \frac{H}{2}e^{\frac{2\pi z}{L}}\cos\theta$
8. Subsurface pressure	$p = \rho g(\eta - z)$	$p = \rho g \eta \frac{\cosh[2\pi(z+d)/L]}{\cosh(2\pi d/L)} - \rho g z$	$p = \rho g \eta e^{\frac{2\pi z}{L}} - \rho g z$

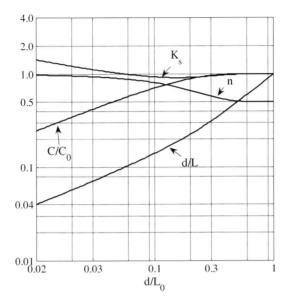

Fig. 1.3 Properties of small amplitude waves.

1.2.2 *Refraction*

It was assumed that energy is transmitted only in the direction of wave propagation under the phenomenon of "shoaling". However, due to large variations in the seabed contours, the waves will start bending as the portion of the wave crest in deeper waters will move faster than the portion of the same crest in shallower waters. This phenomena defined as wave refraction is illustrated in Fig. 1.4.

We can consider a point on a wave front in the offshore and traverse this point towards the shore, which basically represents the wave direction. If we identify yet another point on the same wave front in the offshore separated by a distance, b_0 and trace its movement, it obviously will be separated by a distance "b", the relationship between b_0 and b that could be derived from the famous Snell's law will be governing the behaviour of the waves in the near shore. Equating the power in deep and shallower waters along with the respective directions of the wave, the relationship between the wave heights in the offshore and shallow waters can be derived as

$$\frac{H}{H_0} = \sqrt{\frac{1}{2}\left(\frac{1}{n}\right)\left(\frac{C_0}{C}\right)} \cdot \sqrt{\frac{b_0}{b}} \qquad (1.2)$$

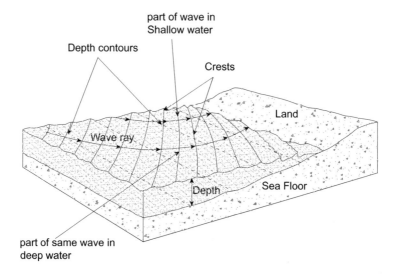

Fig. 1.4   Phenomena of wave refraction.

Herein, $\sqrt{b_0}/b$ is the refraction coefficient will be unity for constant water depth. If it is greater than unity along a certain stretch of the coast, the energy gets amplified leading to erosion, whereas, if it is less than unity, that certain stretch of the shoreline is expected to advance towards the ocean.

In the planning stage of any coastal development program along a stretch of a coast, it is usual to construct a refraction diagram that superposes the bathymetry and the direction of the wavefront from offshore to the shore for a particular predominant wave height, period and direction. A typical refraction diagram shown in Fig. 1.5 provide information on the zones of erosion and deposition of sediments which is vital for the coastal development.

1.2.3   *Diffraction*

When the energy from water waves is transferred laterally along the wave crest i.e., normal to the wave direction, the phenomenon is called wave diffraction. It is dominant around natural barriers or man-made structures such as breakwaters, groins, training walls, etc. The waves curve around the barrier and penetrate into the sheltered area. Sample wave diffraction patterns for a single and a pair of breakwaters are illustrated in Figs. 1.6(a) and 1.6(b) respectively.

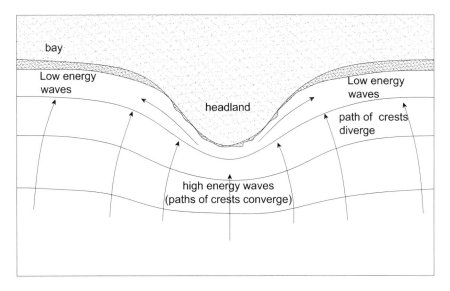

Fig. 1.5   Typical refraction diagram.

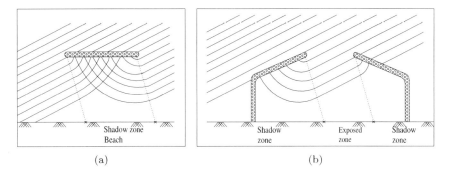

Fig. 1.6   Diffraction of waves around (a) single breakwater (b) a pair of breakwaters.

1.2.4   Breaking

The waves can break either in the offshore in which case the breaking is steepness limited or in the nearshore in which case it is "depth limited". Transfer of excess of energy from wind to the ocean surface in the offshore facilitate breaking of waves. The waves when propagating from offshore to the coast undergo the near shore phenomena like, shoaling refraction or diffraction or a combination of these phenomena. When the waves reach the near shore, it usually steepen while its length reduces resulting in the waves breaking. The most widely adopted criteria for the wave to break are listed below.

(1) When horizontal particle velocity at the crest exceeds the celerity of the wave.
(2) When vertical particle acceleration is greater than acceleration due to gravity.
(3) When crest angle is less than 120°.
(4) When the wave steepness, $H/L > 0.142 \tanh kd$ and for deep waters $\tanh kd$ will become 1.
(5) When wave height is greater than $0.78d$.

The characteristics of the beaches depend also on the type of wave breaking, which in turn, depend on the wave steepness and the beach slope. It is characterised as per the parameter, $N_1 = \tan \beta / \sqrt{H_0/L_0}$. The breaker types namely spilling, plunging, surging and collapsing dictates the extent of mixing of sediments in the breaker zone.

When waves of low steepness waves break over beaches of mild slopes, spilling breakers are generated in which case N_1 is less than 0.5. Breaking is gradual and is by continuous spilling of foam down the front face sometimes called as "white water". Such breakers are expected to move fine sediments **on to the beach**. Plunging breakers occur when waves of medium steep break over beach of medium steepness. The waves at breaking curl over. The breaking is instant and $N_1 = 0.5$ to 3.3. The beaches are characterized by well mixed sediments as they are churned due to the turbulence induced during breaking. Surging breakers occur with the steepest waves breaking over steep slopes. The base of the wave surges up the beach generating considerable foam. It builds up as if to form the plunging type. For this type of breaker $N_1 > 3.3$. A fourth category of breaking waves called as "Collapsing breakers" that would be in between the plunging and surging breakers.

1.2.5 Reflection

When waves strike vertical impermeable obstructions its amplitude will be twice (standing waves) inducing larger forces as well as a large velocity gradient along the vertical plane. This velocity gradient in turn will accelerate scour at the toe of the structure and may lead to its instability. This could be avoided by designing suitable toe protection. It is wise to avoid such vertical face structures inside harbor basins to be exposed to waves as the tranquility will be of great concern and more so at locations exposed to long waves. In most problems in the marine environment, partial reflection occurs due to the absorption of incident or diffracted waves to an extent. An illustration on the reflection and diffraction patterns due to the presence of impermeable barrier is shown in Fig. 1.7.

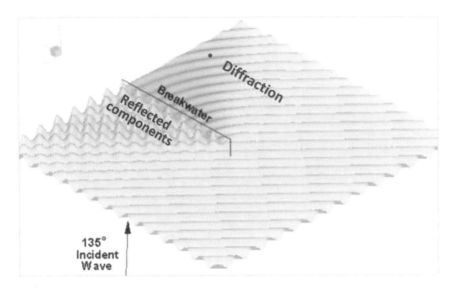

Fig. 1.7 Diffraction and reflection from a barrier.

1.3 The Indian Coast

1.3.1 General

The coastline length of India is about 7500 km, which includes its island territories. The coastline behaviour is not uniform and varies from state to state, even site to site based on local deciding parameters like it orientation, bathymetry, existence of coastal structures, etc. There are nine maritime states along the east and west-coastal regions of India. A wide variation in the geomorphologic features of the coast has been seen with a distinguished features along east- and west-coasts. Further, the northern coastal stretches have unique characteristics with dominant tidal action compared to southern parts of the coasts. The coastal regions of India are densely populated and nearly 20% of the total population of India live in these regions. Further, there is an increase in demand on coastal regions in the recent years for shipping, setting up of industries, developing recreation centers, land reclamation and utilizing marine resources of various kinds. The exploration of natural living and non-living resources in the ocean has necessitated construction of a variety of structures like jetties, dykes, seawalls, groins, platforms, pipelines, etc. which are linked to the economy of the coastal states and ultimately to the national economy.

The west coast of India is characterised by

(i) Stretches from Rann of Katch to Kanyakumari as a *narrow strip* between Arabian Sea and the Western Ghats.
(ii) It has a wider continental shelf
(iii) Lagoons and estuaries are common
(iv) Submerging coast and less attacked by cyclones
(v) Experience both longshore as well as cross shore sediment movement and certain stretches being stable.

The east coast of India is characterised by

(vi) Stretches from Kanyakumari to Sunderbans as a relatively *broader (than western coast)* between Eastern Ghats and the Bay of Bengal.
(vii) The continental shelf is narrow compared to the west coast.
(viii) The coast experiences a subtropical monsoonal climate with an annual rainfall of 1600–1800 mm and severe cyclonic storms.
(ix) Deltas are common.
(x) Most of the major rivers flow into Bay of Bengal leading to a large imbalance in the sediment supply along the coast.
(xi) Longshore sediment transport is dominant.
(xii) Is an emergent coast and more frequented by cyclones.

A considerable portion of the India's coast is affected by coastal erosion, a common phenomenon, by which, a substantial portion of land is taken away by the sea. The causes of coastal erosion include that of both natural and manmade. The Indian coastal zone has come under tremendous pressure due to high population density, industries, tourism, fishing etc. Hence, there is a need to establish the conceptual framework to predict the shoreline and morphological changes for effective usage and protection of coastal zones. Coastal protection structures are an integral part coastal management plans. Several coastal protection structures are proposed and built in the continuous fight against the fury of waves from eating up critical sections of coastlines. Such designs are based on traditional theories supported by design guidelines and software. However, the actual functional performance of the structures during the post construction period is not reported clearly. Hence, it is possible that crucial scientific knowledge is bypassed. This could lead to uneconomical execution of future works and over/under design of future shore protection.

The east coast of Indian peninsula experiences one of the world's highest sediment transport rate, ranging between 0.1 and 1.0 million m^3 per year. The coastline behaviour is non-uniform and varies depending on several parameters such as its orientation, bathymetry and existence of coastal structures. Hence, prior to the planning of the coastal protection measures, it becomes essential to have a clear understanding on the annual wave climate and the existing geomorphology of the coast with the modes of sediment transport and its magnitude through actual measurements in the field. The seasonal wave characteristics prevailing along the study area is most important, as this dictates coastal process and the type of coastal protection measure, based on direction and magnitude of sediment transport. This will also facilitate the calibration and validation of numerical models which are rather scanty or unavailable.

Among the various coastal protection structures, the details of which are discussed later, groins play a major part in protection of coastline, where, longshore sediment transport is dominant. Due to dominance of littoral drift along the east coast of India and to a certain extent along the west coast of India, it is very likely that groin fields and offshore detached breakwaters would continue to be the most important protection measures.

The computation of a reliable longshore sediment transport remains of considerable practical importance in coastal engineering applications. Although, the formulation of Komar (1976) formed the basis for the evaluation of the longshore transport the most commonly adopted method known as the CERC formula (Shore Protection Manual, US Army Corps of Engineers, 1984) which equates the longshore component of wave energy flux entering the surf zone to the immersed weight of sand moved. The effects of particle diameter and bed slope have been included in the estimation of LST by Kamphuis (1991). The estimates of the littoral transport rate for the Indian coast was carried out by Chandramohan and Nayak (1991), and numerous experimental and numerical studies supported by field investigations have been carried out in the past all over the world on the littoral movement. Along the east coast of India, the maritime state of Tamil Nadu is dominated by one of the world's highest littoral to an extent of about

0.8×10^6 m³/annum posing severe erosion problems. Along the west coast, the maritime state of Kerala is noted for its backwaters, development of a number of fishing harbours is exposed to both cross-shore as well a alongshore sediment transport. These were also the only two coastal states of the Indian peninsula that was severely affected by the great Indian Ocean tsunami of 2004 and hence, the authors involved in most of the studies duing the post tsunami period share their experiences on the coastal engineering practices applied to these two maritime states, for which a brief introduction is provided in the next two sections.

1.3.2 Tamil Nadu

Tamil Nadu situated on the south-east of Peninsular India has a land area of nearly 130,000 km² and a coastline of nearly 950 km. The latitude and longitude of all the locations mentioned herein are provided in the location map of Tamil Nadu as shown in Fig. 1.8. A major portion of this coastline, nearly 900 km, which starts from Pulicat in the North and extends up to Kanyakumari in the south, lies on the east bordering the Bay of Bengal. On the western side bordering the Arabian Sea, the coastline is about 40 km between Kanyakumari and Erayumanthurai. Estuaries of ecological importance, major and minor ports, fishing harbours, monuments of international heritage, tourist locations, pilgrimage centres, etc., are located along the coastline of Tamil Nadu. Considerable length of Tamil Nadu coast is exposed to erosion and accretion. For instance, nearly 6 km of coast north of Chennai harbour in Royapuram is getting eroded continuously since the development of the Chennai harbour formed by a pair of breakwaters and nearly about 484,000 m² of coastal area is believed to have eroded along this coast over two decades. It is also a fact that about 225,000 m² of sand has deposited resulting in the advancement of the marina beach towards the sea. The erosion of beaches due to long shore transport has been observed from Pulicat up to Cuddalore. From Poompuhar to Nagapattinam the effect of onshore-offshore movement during cyclone is seen apart from that due to littoral drift. The beaches along the Arabian Sea erode during south west monsoon months and subsequently recover during the non-monsoon months.

The entire coast of Tamil Nadu consists of alluvium and beach sands overlying sedimentary formation such as laterite, limestones, clay, and stones etc. The nature of the coastal belt is as detailed in Table 1.2. The geomorphological features of the coastal regions of Tamil Nadu, extending from Pulicat on the eastern coast to Erayumanthurai on the western coast, vary significantly consisting of sandy beaches, calcareous reefs, bays, tidal inlets, head lands and mangrove marshes.

In the coastal stretch between Chennai and Mahabalipuram beaches on the onshore side are moderate to steep and hence the backshore of beaches is elevated much above the mean sea level. The beaches are formed by the interception of net northerly longshore sand transport. The classical example is the formation of marina beach due to the existence of breakwaters of Chennai harbour. The longshore sediment transport along this stretch has also resulted in closure of river mouths. Between the stretch of the coast from Cuddalore and Nagapattinam the backshore areas have very low relief and fringed with several tidal inlets. For example, Rivers Gadilum and Vellar in their delta regions form several tidal inlets between Cuddalore and Poombuhar before reaching the Bay of Bengal. Similarly the river Cauvary in the delta region between Poombuhar and Nagapatinam branches off into number of distributaries forming several tidal inlets. From the Point Calimere to Tuticorin beaches are wide and have mild slope. The sea coast in this region is exposed to wave action from the Palk Bay and the Gulf of Mannar. The Palk bay is enclosed on three sides by land and has a narrow shallow opening ($\sim$60 km) on the north to the Bay of Bengal. The water depth in the Palk bay is less than 10 m and the bay area is nearly 150 km^2. The fetch length available for wave generation in the Palk bay is small (about 10 km in the North-South direction and 15 km in the East-West direction), and consequently the wave heights observed along its coast are in general small throughout the year. Hence, the wave action along the coast between the Point Calimere and Ramesvaram Island will not be significant. Similarly the coast between Ramesvaram and Tuticorin bordering the northern part of the Gulf of Mannar, which lies between Sri Lanka and India, is not subjected to significant wave action. In the coastal region around Tuticorin, and between Thiruchendur and Kannyakumari the offshore reefs exist nearly for about 500 m from the coast and function like a discontinuous submerged breakwater and protect the coast from the impacts of waves.

On the western side of Tamil Nadu, the coastline facing the Arabian Sea extends from Kannyakumari to Eraiyummanthurai. This coast is steep and has many rocky outcrops. The coast experiences high swell waves during the south west monsoon and these waves breaking near the coast lead to heavy erosion of beaches, and after monsoon accretion takes place and the beaches recover back. Thus, the geomorphology of the coast varies from north to southern tip of the state.

Table 1.2. Nature of coast of Tamil Nadu.

Chennai to Marakkanam	Crystalline rocks overlaid by sedimentary and alluvial formation
Marakkanam to Coleroon mouth	Sand stone, shells, lime stone and clays
Coleroon to Ramanathapuram	Alluvial formation of beach sands and sand dunes that rest on crystalline rocks
Ramanathapuram to Kannyakumari	Alluvial formation of beach sands and sand dunes resting on crystalline rocks
Kannyakumari to Kollengode	Sand and rock

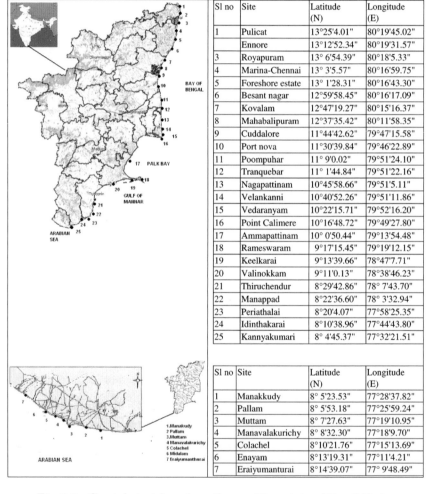

Sl no	Site	Latitude (N)	Longitude (E)
1	Pulicat	13°25'4.01"	80°19'45.02"
2	Ennore	13°12'52.34"	80°19'31.57"
3	Royapuram	13° 6'54.39"	80°18'5.33"
4	Marina-Chennai	13° 3'5.57"	80°16'59.75"
5	Foreshore estate	13° 1'28.31"	80°16'43.30"
6	Besant nagar	12°59'58.45"	80°16'17.09"
7	Kovalam	12°47'19.27"	80°15'16.37"
8	Mahabalipuram	12°37'35.42"	80°11'58.35"
9	Cuddalore	11°44'42.62"	79°47'15.58"
10	Port nova	11°30'39.84"	79°46'22.89"
11	Poompuhar	11° 9'0.02"	79°51'24.10"
12	Tranquebar	11° 1'44.84"	79°51'22.16"
13	Nagapattinam	10°45'58.66"	79°51'5.11"
14	Velankanni	10°40'52.26"	79°51'11.86"
15	Vedaranyam	10°22'15.71"	79°52'16.20"
16	Point Calimere	10°16'48.72"	79°49'27.80"
17	Ammapattinam	10° 0'50.44"	79°13'54.48"
18	Rameswaram	9°17'15.45"	79°19'12.15"
19	Keelkarai	9°13'39.66"	78°47'7.71"
20	Valinokkam	9°11'0.13"	78°38'46.23"
21	Thiruchendur	8°29'42.86"	78° 7'43.70"
22	Manappad	8°22'36.60"	78° 3'32.94"
23	Periathalai	8°20'4.07"	77°58'25.35"
24	Idinthakarai	8°10'38.96"	77°44'43.80"
25	Kannyakumari	8° 4'45.37"	77°32'21.51"

Sl no	Site	Latitude (N)	Longitude (E)
1	Manakkudy	8° 5'23.53"	77°28'37.82"
2	Pallam	8° 5'53.18"	77°25'59.24"
3	Muttam	8° 7'27.63"	77°19'10.95"
4	Manavalakurichy	8° 8'32.30"	77°18'9.70"
5	Colachel	8°10'21.76"	77°15'13.69"
6	Enayam	8°13'19.31"	77°11'4.21"
7	Eraiyumanturai	8°14'39.07"	77° 9'48.49"

Fig. 1.8 Coastal stretches along the maritime state of Tamil Nadu.

1.3.3 Kerala

The southern state of Kerala in India is one of the smallest maritime states with an area of about 39,000 km^2 located along the southwest coast of India lying between Latitudes of 8°18′ and 12°48′N and Longitudes of 74°52′ and 77°22′E. The width of the land area varies between 10 km and 120 km with the maximum width in its central region. Even though, the size of the state is small (only 1% of the total land area of India) it has a sizeable coastal stretch of 590 km which is about 7.8% of the total Indian coast line length of 7500 km. The population density in the coastal districts of Trivandrum, Quilon, Ernakulam, Malappuram and Calicut is more than 2000/km^2 and this is quite high or rather alarming compared to the average state density of about 860 km^2 as per the 2011 Census. The district wise coastline of Kerala is shown in Fig. 1.9.

The coastal stretch of Kerala with a length of about 590 km that forms a major part of the south-west coast of India is considered to be of recent origin according to marine geologists. The Kerala coast has sandy beaches to an extent of 80% whereas, muddy flats and rocky coasts constitute the balance 15% and 5%, respectively. There are evidences of both submergent (lakes, backwaters) and emergent features (barrier spits) which can be attributed to long-term changes in sea level, climate, lithology, structure and non-tectonic movements. Since this part of the coast is considered to be tectonically stable, the sea level change observed along the coast is mainly due to glacio-eustatic impact. The coastal zone experiences heavy rainfall (>3000 mm/year) and generally has high temperature (30°C). The coastal tract is highly dissected and the landscape is the result of extensive Tertiary denudation due to tropical climate (Subramanian and Rao, 1984). The largest spit formed along the west coast of India is the 84 km long and 10 km wide spit formed between the Aleppey (Alappuzha) and Cochin coast, separating the Vembanad Lake from the Arabian Sea (Ahmad, 1972).

The shoreline orientation of Kerala coast is generally straight trending NNW-SSE. The coast is unique due to its biodiversity, presence of a number of coastal features (details presented in Table 1.3) like backwaters, lagoons, estuaries, mud banks, barrier islands, coastal bays, coves, sea cliffs (rocky and lateritic), inland canals etc. The absence of deltas is another conspicuous feature which makes it different from the east coast of India. The coastal strips, estuaries, barrier beaches and lagoons along the Kerala coast comprise the lowlands (elevation < 8 m above MSL) of the three physiographic regions identified in Kerala as indicated in the above figure.

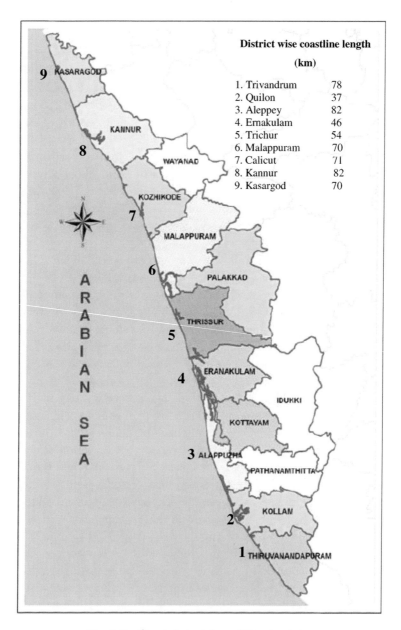

Fig. 1.9 Coastal stretches of Kerala state.

1. Introduction

Table 1.3. Coastal features along the Kerala coast.

Coastal stretch	Location	Geomorphology
Trivandrum-Quilon coast	South Kerala	Rocky and lateritic on crystalline and tertiary formations with alluvial patches. Rocky (Kovalam) and laterite cliffs (Vizhinjam, Varkala) are prominent. Presence of estuaries and lagoons.
Karunagapalli-Calicut	Central Kerala	Sandy coast with number of spits and bars, developed at the mouth of river outlets. Barrier beaches composed of pure alluvium of recent deposits with prominent paleo-strandlines. Presence of estuaries and lagoons. Mud banks present.
Kannur-Kasargod	North Kerala	Sandy coast with number of beach ridges alternating with swales. Wide spread presence of estuaries and lagoons.

There are 44 rivers in Kerala, out of which 41 are flowing towards west and debouch into the Arabian Sea. The remaining three are flowing towards east merging into the backwaters and lakes before entering into the sea. There are five large lakes that lie parallel to the Arabian Sea coast and extend up to nearly half the length of the state. These lakes are linked by canals which in turn are fed by a network of 38 rivers that cross the state. The backwaters of Kerala are in fact the brackish lagoons formed by the action of waves and nearshore currents that create low-lying barrier islands across the mouths of many of the west flowing rivers that originate from the Western Ghats. These backwaters with its unique ecosystem due to the mixing of fresh water from rivers and canals with the sea water, provide a unique environment or habitat for a large number of aquatic flora and fauna which also include some of the endangered species. The picturesque backwaters of Kerala and its beautiful beaches attract a large number of tourists every year.

The mud banks which appear regularly every year at certain locations with the onset of monsoon along the Kerala coast are a unique feature, typical of Kerala. It is a unique phenomenon in which fine grained materials (mostly consisting of silt and clay) are deposited in the inner shelf region of the coast. The length of this formation may vary from 1.5 km to 6 km alongshore, extending a few kilometer (1.5–2) in the offshore direction and are usually elliptical, circular or semi-circular in shape occurring within the 10 m depth contour at a distance of about 5 km to 6 km from the shore.

There are several theories behind the formation of mud banks and its source. In Kerala the source for mud bank could be from river discharges or offshore deposition of fine sediments which are churned up during the initial month of high wave activity during monsoon. There is a significant dampening of the wave activity due to the presence of mud banks and this invariably protects the adjoining coastal area from the fury of monsoon waves. The calm environment provided by mud bank very near to the shore is indeed a boon to the fishermen during the lean monsoon period as the nutrient rich sediment particles in the mud bank attracts lots of fish thereby providing them a safe area for fishing. However, recent studies by various researchers (Moni 1971; Thomas et al. 2013) have pointed out that some of these mud banks are slowly disappearing and a spatial shift in the location along the coast is also observed.

1.4 Summary

An introduction to the behaviour of the ocean waves in the coastal zone, being the main driving force behind the movement of the sediments, which in turn govern the stability of the shoreline have been presented.

Questions that need to be answered through scientific methods are.

(i) Is the coastal stretch considered dominated by onshore-offshore transport or longshore transport?
(ii) What is the magnitude and direction of the sediment transport?
(iii) What are the modifications brought out in the sediment pathways along the coast due to anthropogenic activities (like mining, construction of dams, hard structures like breakwaters, seawalls, groins etc.)?

The above aspects will be discussed in this book with field examples.

References

Ahmad, E. (1972). *Coastal Geomorphology of India*, Orient Longman, New Delhi, 222 p.

CERC (1984). *Shore Protection Manual, Vol. I*, CERC, Dept. of the Army, U.S. Army Corps of Engineers, Washington.

Coastal Engineering Manual (2002). U.S. Army Corps of Engineers, 1110-2-1100, Washington, D.C. (in 6 volumes).

Chandramohan, P. and Nayak, B. U. (1991). Longshore sediment transport along Indian Coast, I.J.M.S. Vol. 20, pp. 114–120.

Kamphuis, J. W. (1991). Alongshore Sediment Transport Rate, *Journal of Waterway, Port, Coastal, and Ocean Engineering*. 117:624–640. doi: 10.1061/(ASCE)0733-950X(1991)117:6(624).

Komar, P. D. (1976). *Beach Processes and Sedimentation*, Prentice Hall.

Moni, N. S. (1971). Study of mud banks along the south-west coast of India, *Proc. Symp. Coastal Erosion and Protection*, KERI, Peechi: 8.1-8.8.

Thomas, K. V., Kurian, N. P., Hameed, T. S. S., Sheela Nair, L. and Reji Srinivas (2013). Shoreline management plan for selected locations along Kerala coast, CESS Project Report, submitted to ICMAM Project Directorate, MoES, Chennai, Vol. 1.

Subramanian, V. and Rao, D. P. (1984). Land use in relation to landforms: A case study of Goa and Karnataka, NRSA Technical Report-0915, 10.

Sundar, V. (2015). *Ocean Wave Mechanics — Applications in Marine Structures*, Wiley. DOI: 10.1002/9781119241652.

Chapter 2

Characteristics and Motion of Sediments

2.1 Introduction

Sediment is any matter in the form of minute separate particles or substance which can be transported by the fluid flow and is ultimately placed as a section of solid elements on the seabed or beneath the surface of a body of water or any other liquid. The process of sedimentation can be simply understood as the deposition of suspended particles by settling. Whereas, mud can be defined as, *"A sediment water mix which consists of materials that are mainly less than 63 μm in size, exhibits poroelastic or viscoelastic rheological behaviour when the mixture is material-supported, and when it is fluid state it is highly viscous and non-Newtonian"*. Very fine particles or sediments in both coastal and inland waters is a matter of widespread attention rising from its importance in the up keeping of navigational channels and in regulating the coastal process. Sediments also influence the coarseness and the resistance of flow due to friction in waterways transport, thereby, leading to the complex problem of stage-discharge-sediment transport relationships.

Oceans comprise of a vast receptacle of land-driven (terrigenous) and comparatively lower proportions materials from volcanic and hydrothermal events (substances thrown out by the volcano). The insoluble materials from the fragmented sediments are left after the breakdown due to chemical failure in bond and breakdown of the pre-existing rocks, which includes silicate minerals, majorly constituting quartz element. The interaction among the atmosphere, lithosphere, and astronomical influence leads to a restless dynamic oceanic system, leading to complex physical and chemical processes; biological action; and the ocean waves, tides, and current circulation. When the ocean waters disperses its energy sufficiently such that it is incapable of keeping the sediments in suspension, results in deposition.

2.2 Sediment Classification

Natural sediments can be classified as into coarse or non-cohesive or cohesionless and fine or cohesive.

Silt and clay are the major constituents of cohesive sediments, the size of which varies up to a fraction of a micrometer. It is noted that often the beds and banks of natural and artificial channels consists of cohesive sediments subjected to erosion, which are highly susceptible to be carried under suspension and eventually deposit at locations along the flow direction.

The basic difference between or non-cohesive and cohesive sediments is that the cohesive sediments can form agglomerations with size, density, and strength much different than those of the original particles under the influence of inter-particle attractive forces. In addition, these properties are not a constant and are bound to be influenced by the forces acting on them during its formation, which in turn can vary with time and quality of pore or ambient water.

Electromagnetic forces in addition to physical drag and lift hydrodynamic forces influence behaviour of cohesive sediments. The movement of non-cohesive open coast marine sediments depends on motion of the sea and physical properties of the sediments. The flow field induced by waves and tides or that due to the gradient formed by salinity, temperature and pressure are the most important driving forces, while extreme events like underwater earthquakes, tsunamis, storm surges will also contribute to the sediment motion. While stating the driving forces, it is mandatory to mention the resistive forces, which are governed mainly by sediment size and characteristics.

2.3 Particle Size

It is the most important parameter, as it partially influences the mode of sediment transport and its corresponding mechanism. Sieve size distribution is one of the most commonly adopted method to represent the particle size by plotting the results as weight of material retained against sieve single size to produce a cumulative size frequency curve. A frequently used classification by coastal engineers for sediment transport calculations is as follows.

2.3.1 *Soil classification*

The classification between coarse and fine sediments is shown in Table 2.1 and is attributed to the mutual interaction of grains in water environment

Table 2.1. Classification based on grain size (Indian Standard Soil Classification System).

Component	Size range
Boulders	>300 mm
Cobbles	80–300 mm
Gravel	
Coarse	20–80 mm
Fine	4.75–20 mm
Sand	
Coarse	2–4.75 mm
Medium	0.425–2 mm
Fine	0.075–0.425 mm
Fines	
Silt	0.002–0.075 mm
clay	<0.002 mm

but not in size. Suspension of coarse grains exhibit independent behaviour from each other, except, for mechanical interactions in highly dense suspensions in the event of coarse sediments constituting the sea bed, only forces of interlocking and friction are to be considered.

In addition, the shape of the grain size curve has an important effect on the properties of sands and gravels. This can be described with two coefficients, the coefficient of curvature C_c and the coefficient of uniformity C_u, defined as follows:

$$C_c = \frac{D_{30}^2}{(D_{60})(D_{10})} \qquad (2.1)$$

$$C_u = \frac{D_{60}}{D_{10}} \qquad (2.2)$$

where

D_{60} = the grain size at which 60% of the soil is finer
D_{30} = the grain size at which 30% of the soil is finer
D_{10} = the grain size at which 10% of the soil is finer.

If C_c is between 1 and 3, the grain size distribution curve will be smooth, and if C_u exceeds 4 for gravels or 6 for sands, there will be a wide range of sizes. When both of these criteria are met, the soil is said to be well graded (designated W); otherwise, it is poorly graded (designated P).

2. Characteristics and Motion of Sediments

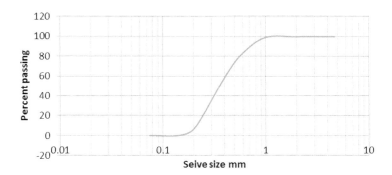

Fig. 2.1 Grain-size distribution curve.

No practical significant can be attached to the shape of the grain size curve for silts and clays.

Figure 2.1 is a typical plot on log normal probability paper for D_{50} and σ the cumulative percent finer is plotted on the probability scale against the grain size on the logarithmic scale. The median particle size may be read at the 50% probability values $[D_{50}]$, which is about 0.35 in the plot below.

Attempts to identify one particular size as characteristic of the whole sample have been made frequently used are median grain size, $D = D_{50}$ geometric mean, D_g grain size corresponding to some fraction such as D_{90}, D_{65}.

The median grain sizes for which 50% by weight of the sample is finer or coarser is denoted by $[D_{50}]$. The diameters D_{90} and D_{65} are the sizes, for which 90% and 65% respectively, by weight of the sample is usually taken as,

$$D_g = (D_{84.1} \, D_{15.9})^{1/2} \qquad (2.3)$$

in which $D_{84.1}$ and $D_{15.9}$ are the grain sizes for which 84.1% and 15.9% by weight of the sediment is finer.

Another classification commonly used by earth scientists is the phi scale.

$$\phi = -\log_2 D_m \qquad (2.4)$$

where, D_m is the grain diameter in mm. The mean value of ϕ for any given sample is usually denoted by M_ϕ.

In addition to the characteristics diameter it is common to specify the nature or spread of the size distribution as follows

$$Skewness = \frac{\log D_g/D}{\sigma_g} \qquad (2.5)$$

$$2^{nd}\ Skewness = \frac{\log(D_{95}\ D_5/D^2)^{1/2}}{\sigma_g} \qquad (2.6)$$

$$Kurtosis = \log \frac{(D_{16}\ D_{95}/D_5\ D_{84})^{1/2}}{\sigma_g} \qquad (2.7)$$

where, σ_g is the standard deviation of the log (grain size) distribution of the sample. It is often found that the distribution of grain sizes is approximately log normally distribution. If this is the case, the standard deviation of the log (grain size) distribution is given by

$$\sigma_g = \log\left(\frac{D_{84.1}}{D_{15.9}}\right)^{1/2} \qquad (2.8)$$

A study on the bed sediment characteristics in the vicinity of outer and inner harbour at the Cochin Port (9°58'10"N, 76°14'22"E) with typical values is presented. The locations around the outer and inner harbour of port are depicted in Fig. 2.2.

The bed sediment characteristics established for a few locations of outer and inner harbour of Cochin port are tabulated in Table 2.2.

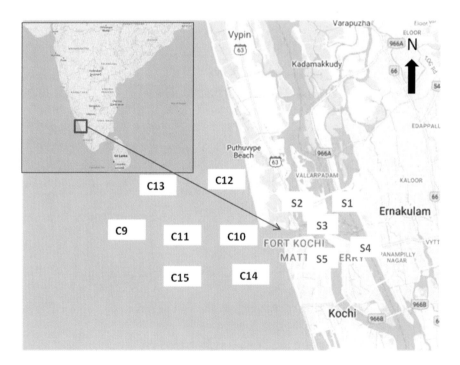

Fig. 2.2 Sediment sample locations.

Table 2.2. Bed sediment characteristics.

Locations	D_{10}	D_{30}	D_{60}	C_u	C_c
$S1$	0.21	0.3	0.4	1.9	1.07
$S2$	0.2	0.26	0.38	1.9	0.88
$S3$	0.19	0.41	0.52	2.73	1.7
$S4$	0.17	0.27	0.5	2.94	0.85
$S5$	0.19	0.3	0.49	2.57	0.96
$C9$	0.19	0.35	0.52	2.73	1.24
$C10$	0.18	0.29	0.5	2.77	0.93
$C11$	0.19	0.38	0.51	2.68	1.49
$C12$	0.21	0.4	0.53	2.52	1.43
$C13$	0.24	0.4	0.53	2.21	1.25
$C14$	0.2	0.32	0.51	2.55	1
$C15$	0.2	0.35	0.5	2.5	1.225

The points $S1$ to $S5$ denote the inner harbour locations and locations $C9$ to $C15$ denote the outer harbour locations. The average value of C_u across the sample points in the inner and outer harbour area is found to be about 2.4 and 2.5. The average value of C_c is about 1.09 and 1.2, from which we can understand that the soil sample is well graded. The C_u and C_c in the inner harbour locations are found to be less compared to the locations in the outer harbour locations.

2.4 Plasticity

It is necessary to have a relatively simple test that reflects the influence of the size and nature of the particles. This is accomplished by the liquid limit (LL) and plastic limit (PL). Tests to determine these limits are highly arbitrary, but they nonetheless have great practical value. The liquid limit is defined as the moisture content at which soil begins to behave as a liquid material and begins to flow. The liquid limit is determined in the lab as the moisture content at which the two sides of a groove formed in soil come together and touch for a distance of 2 inch after 25 blows. Details are available in ASTM [2017].

Atterberg defined the plastic limit of clay as the limit of the moisture content up to which it exhibits plastic property and below it becomes friable or crumbly.

The plasticity index is the difference between the liquid and plastic limits. It denotes the range of water content over which the remolded clay exhibits characteristics of a plastic material.

Thus
$$PI = LL - PL. \qquad (2.9)$$

2.5 Shape

The shape factor of a sediment particle is governed by its geometric shape and is given as,

$$SF = \left(\frac{D_1}{(D_2 D_3)^{1/2}}\right) \qquad (2.10)$$

where D_1, D_2, D_3 are respectively the lengths of the shortest, intermediate and longest mutually perpendicular axes.

2.6 Fall Velocity (v_f)

When the flow is dominated by suspended sediments, its gravitational fall velocity is of great interest. The fall velocity acts as a restoring force against the turbulence entraining forces driving the sediments. Natural sediments in real time are scarcely spherical, although an approximation of fall velocity over rigid sphere is used for theoretical calculations. The fall velocity in fact, dictates the quantity of sediments that can deposit in a channel, like the approach channel of a harbour and its magnitude is directly governs the quantity of maintenance dredging. In other words, the fall velocity is the terminal velocity attained by an isolated solid grain settling due to gravity in a still, unbounded, less dense fluid.

The fall velocity, v_f is the final equilibrium velocity reached by the falling sphere. Under these circumstances the drag of the fluid must exactly balance the force due to gravity tending to pull the sphere down.

$$\frac{\pi D^3}{6}(\rho_s - \rho)g = C_D \frac{\pi D^2}{4} \rho \frac{v_f^2}{2} \qquad (2.11)$$

C_D: Drag Coefficient.

Figure 2.3 shows the typical variation of C_D with Reynolds No: $R_e = \frac{v_f D}{\nu}$ for a sphere in an infinite fluid. In the Stokes region, that is, for $\frac{v_f D}{\nu} < 0.1$

$$C_D = \frac{24}{(v_f D/\nu)} \qquad (2.12)$$

$$v_f = \frac{gD^2}{18\nu}\left(\frac{\rho_s - \rho}{\rho}\right) \qquad (2.13)$$

for $B < 39$

$$v_f = \left[\left(\frac{\rho_s - \rho}{\rho}\right)g\right]^{0.7} D_{50}^{1.1}/6\nu^{0.4} \qquad \rightarrow 39 < B < 10^4 \qquad (2.14)$$

$$v_f = \left[\left(\frac{\rho_s - \rho}{\rho}\right)gD_{50}/0.91\right]^{0.5} \qquad \rightarrow 10^4 < B \qquad (2.15)$$

where,

$$B = \left[\frac{\rho_s - \rho}{\rho}\right]gD^3/\nu^2 \qquad (2.16)$$

Figure 2.3 shows that for $400 < R_e < 200{,}000$, the C_D is almost a constant. From Eq. (2.17) this would give

$$v_f = const \times \left(\frac{\rho_s - \rho}{\rho}gD\right)^{1/2} \qquad (2.17)$$

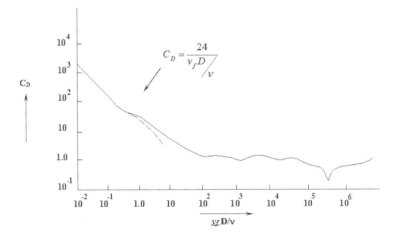

Fig. 2.3 Typical variation of C_D with R_e falling in an infinite fluid.

At 20°C the fall velocity for particles of median size $60 \times 10^{-6} < D_{50} < 6000 \times 10^{-6}$ m is given by

$$Log\left(\frac{1}{v_f}\right) = 0.447(\log D_{50})^2 + 1.961 \log D_{50} + 2.736 \qquad (2.18)$$

The expressions obtained for a sphere are not directly applicable to grains of sediment because of the difference in shape. The effect of variations in shape on the fall velocity is much less significant for small grain sizes than for large. Approximate formulae may be obtained for relatively coarse and relatively fine sediments. For quartz sand in water at 20°C the particle size will be between 3 mm and 90 mm. The fall velocity for different shape factors, SF is given as

$$SF = 1 \quad v_f = 6.5 \ D^{1/2}$$
$$SF = 0.7 \quad v_f = 4.2 \ D^{1/2}$$
$$SF = 0.3 \quad v_f = 2.8 \ D^{1/2}$$

For grain size less than 0.1 mm, $v_f = 92 \times 10^4 \ D^2$.
In the above formulae, v_f is in m/s and D is in meters.

2.7 Angle of Repose

It is the steepest angle formed by a heap of sediments without losing its slope and is on the verge of collapsing. It is to be noted that the friction force restraints in addition to the inertia, opposing the movement of non-cohesive sediments at contacts. The capacity of a particle to resist sliding motion relative to its submerged gravity component normal to the sliding is expressed as the friction coefficient μ_d; therefore it represents the ratio of the tangential resistive force to the downward normal force.

2.8 Effect of Temperature

A change in temperature modifies the coefficient of viscosity of the fluid and hence the Reynolds no: $\frac{v_f D}{\nu}$. It is clear that a change in R_e will have a significant effect on the C_D and hence on the fall velocity at small values of R_e but a relatively small effect in the range $400 < R_e < 200{,}000$.

Temperature also has an effect on the density of the water. However, the change in density is so small compared with the change in ν that density variation may be neglected for all practical purposes.

2.9 Effect of Sediment Concentration

A small cloud of grains in an otherwise clear fluid will settle faster than a single grain.

Richardson and Jeronimo [1979] suggested

$$\frac{v_{f_c}}{v_{f_0}} = (1-C)^n \qquad (2.19)$$

2. Characteristics and Motion of Sediments

where

v_{f_c} = is the fall velocity at concentration C.
v_{f_0} = is the fall velocity in clear fluid.
$n = 4.6$ for low R_e and $n = 2.3$ for high R_e.

In some fluids, particles of silt and clay tend to collect together in flocs is many times greater than that of individual particles so that under these circumstances, an initially uniform suspension of particles may produce accelerated settling.

2.10 Effect of Turbulence

At very small R_e, drag exerted by the fluid on a body in a stream is directly proportional to the relative velocity. Thus, random fluctuations in velocity will have no effect on the mean drag, which will remain proportional to the mean velocity. On the other hand, in the region where C_D is approximately constant, drag is proportional to the square of the velocity. Under these circumstances the mean drag will no longer be proportional to the mean velocity if random fluctuations are present.

2.11 Permeability and Porosity

Permeability is the property of the soil to transmit water and air through its pores. The higher the degree of permeability, the greater will be the extent of seepage. The permeability of a bed of sediments is usually defined in terms of the flow produced by a given pressure gradient.

For different soil types as per grain size, the orders of magnitude for permeability are shown in Table 2.3.

Table 2.3. Permeability for different soil types.

Soil types	Permeability
Gravel	10^0 cm/s
Coarse sand	10^0 to 10^{-1} cm/s
Medium sand	10^{-1} to 10^{-2} cm/s
Fine sand	10^{-2} to 10^{-3} cm/s
Silty sand	10^{-3} to 10^{-4} cm/s
Silt	1×10^{-5} cm/s
Clay	10^{-7} to 10^{-9} cm/s

Most wildly used formula for permeability after Fair and Hatch [1933]

$$k = \frac{Kg}{\nu} \qquad (2.20)$$

where,

$$K = \frac{1}{A\left[\frac{(1-n)^2}{n^3}\left(B\sum \frac{P}{100D_{gm}}\right)^2\right]} \qquad (2.21)$$

where K = specific permeability, k = coefficient of permeability, n = porosity, A = packing factor (taking a value of about 5), b = sand shape factor ranging between 6.0 to 7.7 for spherical grains to angular grains respectively, P = percentage of sand held between two adjacent sieves, and D_{gm} is the geometric mean of the mesh sizes of the two sieves.

$$n = \text{Porosity} = \frac{\text{volume of voids in a sample of sediment}}{\text{Total volume of the sample}}$$

2.12 Bulk Creep

Continuous yielding of soil particles in un-drained stress conditions is termed as creep or plastic flow. Attributable to the repeated loading effect, creep exhibits a long term depreciation/weakening in the shear strength of the sediments. Recurring load reversals are susceptible to occur during earthquakes and with much lesser impact during its exposure to continuous wave action. Cyclic stress level, periodicity, frequency and duration, and sediment types are the factors on which the effect of repeated loading will depend.

The pore water pressure will tend to increase along with the stress, when a recurring load is applied on soils with poor drainage and fine sands in a confined space. This leads to reduction in shear strength of the residue over time and catastrophic events like earthquake and stormy wave increases the fluid pressure which in turn results in further reduction of shear strength. For huge land mass, this reduction in shear strength influenced by pore water pressure may lead to liquefaction.

2.13 Sediment Motion

2.13.1 *General*

The motion of sediment such as erosion or deposition on the seabed is normally termed as fluvial process. Erosion can happen either when the flow of water over seabed influences the direct shear stress on the bed, or when the seabed is susceptible to such stresses due to fine and loosely movable sediments. In the event of ocean waves transmitting substantial amounts of sediments, it can have the propensity to control the wear of seabed (abrasion). Meanwhile, the abraded sediments are crushed down, thereby, reducing in size and more smoothed (attrition). Although sediment settles down in a slow moving or still water in ponds or lakes and ocean, river channel deposits and beach sands are some of examples of fluvial transport and deposition.

Sediments in ocean is transported either as bed load which usually is coarser or suspended load which usually is finer. The velocity of the flow reaches a critical value for initiating the motion of the sediments which depend on it grain size. This is referred to as the entrainment velocity. Nevertheless, even if the velocity falls below the entrainment velocity the grains will continue to move due to the reduced (or removed) friction between the grains and the river bed. Ultimately, when the flow velocity reduces, in particular falls below its threshold or if the grain size is large the sediments finds its way to the river or seabed.

Sands from medium to coarse and clays which are non-sensitive will typically move as there or thereabouts firm material for which the velocities are fairly low. Causes of this volume will tend to give up relatively rapidly when the side angle is cut down under some critical point.

Edgers and Karlsrud [1982] proposed three potential mechanisms for these huge run-out. The mechanisms include (1) turbidity currents, (2) viscous flow models, and (3) progressive liquefaction.

2.13.2 *Incipient sediment motion*

Laboratory experiments conducted extensively indicates two criteria for movement initiation of level bed sediment with D_{50} grain size distribution between 0.1 mm and 0.2 mm [Hallermeier, 1980]. The appropriate threshold flow velocity for sand motion as applied in the field is

$$u_{\max(-d)} = \left[8 \left(\frac{\gamma_s}{\gamma} - 1 \right) g D_{50} \right]^{0.5} \qquad (2.22)$$

where

$u_{\max(-d)}$ is peak fluid velocity at the sediment bed.
$u_{\max(-d)}$ can be determined using Airy's wave theory given as

$$u = \frac{\pi H}{T} \frac{\cosh k(d+z)}{\sinh kd} \sin(kx - \sigma t) \qquad (2.23)$$

Hence

$$u_{\max(-d)} = \frac{\pi H}{T} \frac{1}{\sinh kd} \qquad (2.24)$$

$$\frac{T u_{\max(-d)}}{H} = \frac{\pi}{\sinh kd} \qquad (2.25)$$

2.14 Liquefaction of Sands

2.14.1 General

When saturated sand is subjected to loading without drainage, the pressure in the pore fluid may, under certain circumstances, approach or equal the total stress to which the sand is subjected. The total stress σ minus the pore pressure, p_p is termed the effective stress σ', or

$$\sigma' = \sigma - p_p \qquad (2.26)$$

Under such condition, the effective stress approaches 0. Shear strength and deformation of soils are controlled by the effective stress; as it approaches 0, the shear strength likewise approaches 0. At this point the sand becomes fluidized or liquefaction occurs and the sand is unable to support loads.

Under cyclic loading conditions, soils will experience an increase in strain with each load application. Laboratory studies have shown that the pore pressures in saturated, un-drained sands build up progressively under the action of these cyclic stresses. This is true for even dense sands, although in dense sands, the magnitude of an increase in pore pressure with each stress reversal may be quite small. However, with a sufficient number of stress reversals, liquefaction may occur, provided the excess pore water pressures do not dissipate between load applications.

Cyclic stresses produced by waves can also induce pore pressures in marine sediments, and some evidence of liquefaction under such conditions has been observed. At any instant, the pore pressures in sediment will depend on the state of denseness, the permeability of the sediment, and the induced cyclic stress ratios resulting from the waves. Permeability is important because waves have much longer periods than earthquakes, and the pore pressures can change due to redistribution within the sediment as well as dissipation at free-draining boundaries. Because the problem is one of partial drainage rather than no drainage, it is more complicated than the undrained earthquake problem.

2.14.2 Soil types susceptible to liquefaction

Most liquefaction studies have centered on sands as the liquefiable material. Again, this is probably the direct result of earthquake-induced liquefaction studies. Earthquake shocks are of high frequency and short duration, and they can induce high stresses in the soils. Sands are the materials, which seem most readily to liquefy under these conditions. Sands, of course, are cohesion less materials, that is, they have no shear strength at zero effective stress, and it is this property, which allows liquefaction to occur. Sediments which are predominantly silt with little or no clay are also cohesion less; this type of material was involved in the problems mentioned earlier regarding the 3 m diameter pipe. Generally, clays are not susceptible to liquefaction, although the Mississippi Delta material, which lost strength during storms to the point that instruments sank, was clay; however, it was extremely weak (un-drained shear strength of approximately $366 \, \text{kg/m}^2$).

By and large, sands and coarse silts should be suspected of being liquefaction-prone materials. The in situ state of denseness must also be considered in the determination of liquefaction. Loose materials, especially those in a metastable particle arrangement, will liquefy more readily than dense ones, but if a storm is long enough (with a large enough number of wave-induced stress cycles), even dense materials are capable of liquefying. [It is emphasized that the analyses and test methods, which follow, rather than empirical methods, should determine liquefaction potential, since most empirical methods were developed for earthquake analyses].

2.15 Shields Curve

It was shown by Shields [1936] that critical shear stress value could be expressed as a function of the Reynolds number.

$$\frac{\tau_{cr}}{(\rho_s - \rho)gD_{50}} = f(R_e) \tag{2.27}$$

where,

R_e: Reynolds number $= v_* D/\nu$
τ_{cr}: critical shear stress
ν: kinematic viscosity
v_*: shear stress velocity.

From Eq. (2.27), the critical shear stress (τ_{cr}) can be found by an iterative process. In Fig. 2.4 lines are drawn indicating where, the above equation holds for one particular grain diameter. With the help of these lines it is

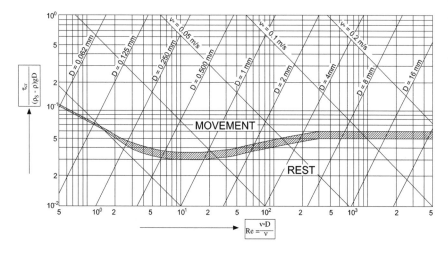

Fig. 2.4 Shield's curve (the hjulstorm approach).

easy to find the critical value (at the intersection of the line and the curve; since the limits are not sharp, the "curve" is indicated by a hatched band).

As the original shields curve was difficult to use because of the presence of same variables in both axes, the graph became implicit. In order to make the graph explicit, many researchers created equations to approximate shield's curve. The graph was transformed to another axis system with dimensionless grain diameter D^*. Figure 2.5 shows the shield's parameter as a function of dimensionless parameter.

$$D^* = D_{50} \left[\frac{(s-1)g}{v^2} \right]^{1/3} \qquad (2.28)$$

where $s = \frac{\rho_s}{\rho}$.

Whether or not there is any movement of grains thus depends on the parameters of velocity, V (or critical shear stress, τ_{cr}) and the diameter, D. These three parameters are taken into account in nearly all sediment transport theories.

Several features of the Shields diagram are particularly noteworthy.

The Shields stress is a minimum for sand whose grain diameter is in the range of (0.06–2.0 mm). Sand is small enough to have small mass but too large for adhesion forces to come into play.

- Silt/clay, in spite of the smaller size, requires a higher shear stress for motion than sand. Here adhesion forces become overwhelmingly large

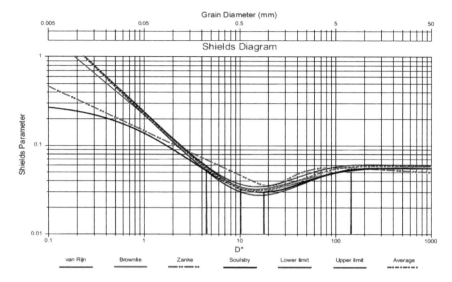

Fig. 2.5   Shields parameter as function of dimensionless parameter.

and bind the sediment together into a mass that is very resistant to erosion.
- The Shields parameter for gravel is constant at 0.06, implying that Shields stress here becomes a simple function of grain size.

Problem I

Quartz sediment in seawater with a median sediment dia $D_{50} = 0.15$ mm. Find,

(a) With wave period $T = 10$ s, the minimum wave height for sand motion in water depth $d = 10$ m.
(b) With wave period $T = 8$ s, the maximum water depth for sand motion with $H = 2$ m.
(c) with $H = 1$ m, $d = 20$ m the minimum wave period for sand motion $D_{50} = 0.15$ mm $= 0.00015$ m.

Solution

$$u_{\max(-d)} = \left[8\left(\frac{\gamma_s}{\gamma} - 1\right)gD_{50}\right]^{0.5} = \left[8\left(\frac{2.65}{1.026} - 1\right)9.81 \times 0.00015\right]^{0.5}$$
$$= 0.1365 \,\text{m/s}$$

(a) For $d = 10\,\text{m}$, $T = 10\,\text{sec}$, $L_0 = 156$, $d/L_0 = 0.0641$, $\therefore \sinh kd = 0.74$

$$\frac{u_{\max}(-d)T}{H} = \frac{\pi}{\sinh\left(\frac{2\pi d}{L}\right)} = \frac{\pi}{0.74} = 4.4$$

$$\therefore H = \frac{u_{\max}(-d)T}{4.4} = \frac{0.1365 \times 10}{4.4} = 0.32\,\text{m}$$

(b) $T = 8$, $H = 2$, $d = ?$

$$\frac{u_{\max}(-d)T}{H} = \frac{0.1365 \times 8}{2} = 0.546$$

we know, $\dfrac{u_{\max}(-d)T}{H} = \dfrac{\pi}{\sinh kd}$

$$0.546 = \frac{\pi}{\sinh kd}$$

$\sinh kd = 5.75$

Corresponding

$$\frac{d}{L_0} \approx 0.3840$$

$$L_0 = 1.56 \times 8^2 = 99.84$$

$$\therefore d \approx 40\,\text{m}$$

(c) $H = 1$, $d = 20\,\text{m}$, $T = ?$

$$\frac{u_{\max}(-d)T}{H} = \frac{\pi}{\sinh kd}$$

$$u_{\max(-d)} = \left[8\left(\frac{\gamma_s}{\gamma} - 1\right)gD_{50}\right]^{0.5}$$

$$= 0.135$$

$$\frac{u_{\max}T}{H} \quad \text{or} \quad \frac{u_{\max(-d)}T}{H} = 0.135$$

$T = 6.6\,\text{s}$

$u_{\max(-d)} = 0.18\,\text{m/s}$

which is somewhat larger than the threshold velocity, For $T = 5$,

$$\frac{u_{\max(-d)}T}{H} = 0.25, \quad u_{\max(-d)} = 0.05\,\text{m/s}$$

which is much less than the required threshold.

Hence, by trial and error $T = 6.6\,\text{s}$.

Problem II

What shear stress will move a 2.0 cm particle resting on the seabed?

Solution

From the graph shown in Fig. 2.5, for the given grain size particle (read from horizontal axis), the line intersecting the curve corresponds to the shield's parameter (read from vertical axis).

Shields parameter = 0.05

$$\frac{\tau_{cr}}{(\rho_s - \rho)gD} = 0.05$$

$$\tau_{cr} = 0.05 \times (2650 - 1025) \times 9.81 \times 0.02$$

$$\tau_{cr} = 15.94\,\text{Nm}^{-2}$$

The shear stress of $15.94\,\text{Nm}^{-2}$ will initiate the movement of a particle of diameter 2.0 cm.

References

ASTM D4318-17, Standard Test Methods for Liquid Limit, Plastic Limit, and Plasticity Index of Soils, ASTM International, West Conshohocken, PA, 2017.

Edgers, L. and Karlsrud, K. (1982). Soil flow generated by submarine slides — Case studies and consequences, *Proc. 3rd Int. Conf. Behaviours of Offshore Structures*, pp. 425–437.

Fair, G. M. and Hatch, L. P. (1933). Fundamental factors governing the streamline flow of water through sand, *J. Amer. Water Works Assoc.*, 25, 1151–1565.

Hallermeier, R. J. (1980). Sand motion initiation by water waves: Two asymptotes, *Journal of Waterway, Port, Coastal and Ocean Engineering*, 106, No. 3, 299–318.

Richardson, J. F. and Jeronimo, M. A. D. (1979). Velocity-voidage relations for sedimentation and fluidization, *Chem. Eng. Sci.*, 34, 1419–1422.

Shields, A. (1936). Anwendung der Aehnlichkeitsmechanik und der Turbulenzforschung auf die Geschiebebewegung, *Mitt. Preuss. Versuchsanst. Wasserbau Schiffbau*, 26, 36 pp.

Chapter 3

Sediment Transport

3.1 Introduction

In the coastal and nearshore zones, where shore processes begin, there is mixing, sorting and transportation of sediments as well as run-off from the land. Waves, winds and currents mould the shorelines of the world and their interaction with the land and its run off determines the configuration of coastlines and the adjacent bathymetry. Man's intervention into the natural processes taking place in the coastal zone also has an impact on the shoreline. These are in the form of discharges into the ocean, thermal and radioactive pollution, dredging, coastal construction, mining and poaching.

There are many factors the affect the processes that take place in the nearshore zone. The most importance are: degree of exposure to waves and currents, supply of sediments and run off to the coasts, topography of the continental shelf and the adjacent coast and tidal range and the intensity of the tidal current and coastal climate.

Of all these factors, waves play a predominant role in the nearshore processes.

3.2 Modes of Sediment Transport

3.2.1 *General*

Sediments are considered to be transported by one of the two major modes i.e., either as bed load or suspended load. When the sediments are transported along the seabed either by rolling, sliding or bounding it is said to be bed load sediment transport. When the transport rates are high, which involving movement of bed materials in several layers with all in contact with one another are said to be a sheet flow. The transport by hopping motion is usually referred to as saltation, whereas, the sediment transported by suspension in the moving fluid is said to be suspended load sediment transport. In addition, when the fine particles are carried by moving fluid in suspension, from the river estuaries are said to be wash load.

3.2.2 Description of the threshold of movement

A perfectly round object or sphere will readily move on minimum horizontal force if it is resting on a flat smooth horizontal surface. In the real case, the area of an erodible boundary, the sediment particles are not flawlessly round and it is of non-uniform sizes, on the other hand the seabed is not flat or horizontal and it may be of some slope too. In such a case, the application of horizontal force will lift the sediment particles only when the force is greater than the resisting force by the sediment particles due to such natural resistance. A fluid in motion applies shear force at the point of interface to the sediments. This implies that a considerable amount of force is applied to the top layer of materials. Based on experimental observations by many researchers, it is found that when the shear force is gradually increased from zero, a point is reached at which particle movements can be observed at a number of small areas over the bed. A further small increase in shear force is usually sufficient to generate a widespread sediment motion as the bed load type. This defines the "threshold of motion" and the associated critical shear stress. Figure 3.1 shows the moving fluid applying shear force and the dislodging of prominent grains.

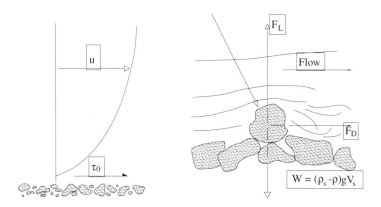

Fig. 3.1 Moving fluid causing sediment movement.

Where, τ_0 — shear force, F_L — lift force, F_D — drag force, ρ_s — density of sand, ρ — density of water, g — acceleration due to gravity and V_s — volume of the particle. Detailed discussion on bed shear stress due to waves, currents and combined action of waves and currents are given in Sec. 6.10.

3.2.3 Load transport general approach

Generally sediment transport takes place either as bed load or the combination of both bed load and suspended load. The transport by suspended load alone is rare and occurs only in the case of fine silts. The combined load transport in general is referred to as total load. In practice, bed loads and suspended loads will be calculated separately and the resultant sum will be the total load. Practically, the separation of bed and suspended load are very difficult, researchers have tackled the total load problem by matching the transport equations at well-defined height.

3.3 Littoral Transport

3.3.1 General

Littoral transport is defined as the movement of sediment particles within the surf zone, that is, the zone close to the shoreline. If the flow involves sediments, it is no longer said to be a simple fluid flow as it involves two components. Sediment transport can be influenced by either waves or currents or a combination of both. The theory of momentum flux is used to explicate the phenomenon of wave set-up and longshore current generation. It is explained as the excess flow of momentum due to the presence of waves (with units of force/unit length). It arises from the orbital motion of individual water particles in the waves. The knowledge of littoral transport gives a clear understanding of momentum flux theory. Sediment transport is defined as the product of instantaneous concentration and instantaneous velocity as given,

$$S = \frac{1}{t'} \int_0^{d+\eta} \int_0^{t'} c(z,t) u(z,t) dt dz \qquad (3.1)$$

where, S — sediment transport rate expressed in [m^3/m^2], t' — integration time, d — water depth, η — instantaneous water surface elevation, $c(z,t)$ — instantaneous concentration of material, $u(z,t)$ — instantaneous velocity component, z — elevation above the sea bed and t — time.

Sediment transport is classified as cross-shore transport and as alongshore transport based on the direction of net drift. If the average net drift of sediments is perpendicular to the shoreline, it is referred to as cross-shore transport, whereas, if the net drift of sediments is parallel to the shoreline, it is said to be alongshore sediment transport. The instantaneous motion of sedimentary particles has both a cross-shore and alongshore component. Both alongshore and cross-shore transport are significant in the surf zone.

In general, if sediment is moved by alongshore transport, it will not return to its initial position. The cross-shore transport moves sediment over the beach profile. The beach profile is adapted only to the momentary wave characteristics.

In the surf zone breaking wave induces high orbital velocities which are responsible for the significant suspended sediment concentration. In addition, the wave-generated longshore current in the surf zone can cause high longshore sediment transports. In the offshore, the same principles are valid as in the surf zone (wave-induced suspended sediment concentrations, longshore sediment transport), however being less intensive.

The wave propagating towards shallow water depth influences the sediments at the bottom when its oscillatory motion begins to feel the sea bed. Initially, only the material of very low density (such as organic and other seaweed matters) gets initiated. These particles oscillate back and forth with the orbital motion of the waves. For a given wave condition in deep water, the wave height increases as the depth decreases towards the shore (to be correct first a slight decrease of the wave height occurs). This increasing wave height and decreasing the depth produce an increasing orbital velocity. At a certain location, the orbital velocity, exert adequate shear to initiate the sediment particles. The sediments then leads to formation of ripples with the crests parallel to the wave crests. These ripples are typically uniform and periodic, and the sand moves from one side of the crest to the other with the passage of each wave. When the waves become even higher and finally break, the orbital velocities continue to increase. Finally, the orbital motion will exceed the critical velocity of sheet flow. At this point, the ripples are flattened. A layer of sediment now moves over the bottom resulting in higher concentration of suspended sediments.

Littoral transport results due to the interaction of waves, currents, tides, sediments, gradients, winds and other process in the surf zone. The present chapter discusses the processes which are involved in sediment motion which can be divided into three categories namely, the forces which cause the longshore current, the sediment sources and the transport mechanism.

The definitions are given in the terms used and the axis-system.

3.4 Definitions

Prior to describing the processes involved in sediment motion, it is necessary to define the terms used in the field of coastal engineering like "beach", "shore", etc. (Fig. 3.2). In the definitions use is made of the "Glossary of Terms" of the Shore Protection Manual [1984].

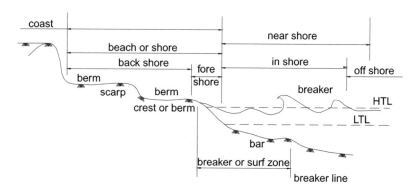

Fig. 3.2 Definition sketch.

A beach is the region of unconsolidated particles that spreads landward from the lowest low water line to the consolidated region where significant changes are seen (start of cliff or dunes). The seaward limit of a beach is the mean low water line (the intersection of any standard low tide datum plane with the shore).

A coast is the stretch of land area from the shoreline up to a permanent feature of the land mass. The area bordered by land and sea is the coastal area. The area extending landward from the sea is Onshore. According to beach terminology, offshore is a moderately even zone of adaptable breadth, starting from the breaker zone to the edge of the continental shelf seaward. The shore is the thin strip of land in direct contact with the sea water, that is, the area between the low tide level and high tide level.

The axis-system used can be described as follows (Fig. 3.3).

The x-axis is horizontal and parallel to the shoreline. It is directed positively to the right for a person standing on the beach facing the ocean. The y-axis is also horizontal, perpendicular to the shoreline and positive in the direction of the sea. Waves approaching with crests parallel to the coast are therefore travelling along the y-axis in the negative direction. In the case of a description of the shoreline the x-y plane is placed in the still water level. The x-axis therefore coincides with the shoreline. For sediment transport considerations the x-y plane is placed on the sediment bed in the water. The z-axis is then vertically upwards from the bottom.

3. Sediment Transport 43

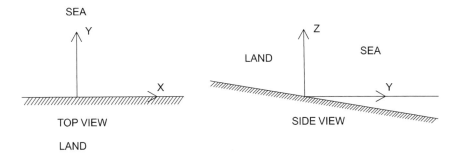

Fig. 3.3 Definition axis system.

3.5 Driving Forces

Water generally moves in some direction or other along the coast. If a total water column is considered, there will be driving forces that will initiate its movement. On the other hand, there are also forces which try to resist the driving forces. The equilibrium of these forces in the longshore direction results in a constant current along the coast (the longshore current). The driving forces will include the radiation stress components, wind forces and the tide which is generally present. The turbulence (horizontal diffusion mechanism) and the bottom friction will constitute the resisting forces.

3.6 Radiation Stresses

The concept of radiation stress and its components play a significant role in coastal morphological processes. The discussion herein is brief and for further detailed descriptions the reader is directed to refer to Longuet-Higgins and Stewart [1960, 1962], Dorrestein [1961] and Battjes [1974].

Waves influence the mean condition (averaged in time) of the medium in which they propagate. The mean effect of the waves can be expressed in extra terms in the mass and the momentum balance of the medium. Due to the change in the mass and momentum balance, the medium behaves differently when waves are absent. The wave-induced contribution of horizontal momentum to the mean balance of momentum is referred to as the radiation stress (Fig. 3.4), introduced by Longuet-Higgins and Stewart [1960].

The change of horizontal momentum is shown in Fig. 3.5. The pressure under propagating waves fluctuates. If we assume that the absolute values of the pressure fluctuations under the wave crest and the wave trough are equal (which highly questionable), then there would still be a net effect caused by the increased pressure acting on an increased area (under the wave crest),

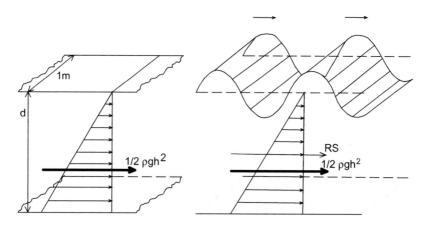

Fig. 3.4   Radiation stress (caused by waves).

and the reduced pressure acting on a reduced area (under wave trough). The mean value of the pressure fluctuations integrated over the depth is not equal to zero, but a positive. This positive value of the integrated pressure fluctuations forms one part of the radiation stress. Another contribution is found by the non-zero integrated velocity fluctuations.

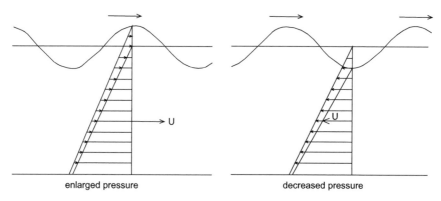

Fig. 3.5   Wave-induced changes in horizontal momentum caused by pressure and velocity fluctuations.

The radiation stress is found to be the wave-induced contribution of horizontal momentum (to the mean balance of momentum). Since, a rate of transfer of momentum is equivalent to a force [second law of Newton: $F\ dt = d\ (mv)$], the change in radiation stress is the wave-induced contribution of force to the water through which the waves propagate. Such forces, applied to a given volume of water, can be a non-zero resultant. In the new balance of momentum, these wave-induced forces have to be balanced which can be achieved by a pressure gradient (slope in the mean water level) or a few velocity (longshore current) with counteracting bottom shear stress.

In reality, the radiation stress is neither a true stress (force per area, N/m^2), nor a true force (N), but a force (force per unit length, N/m) over the entire depth (resulting from the integration of the force per unit area over the water depth).

Unlike hydrostatic pressure, the radiation stress is not isotropic; indeed, just as with stresses, it is associated with a given direction or plane. In this discussion, these planes are vertical and perpendicular to the two horizontal axes, X oriented in the direction of wave propagation and Y along the wave crest. This will yield the principal stresses; S_{XX} and S_{YY}. (Note: when the radiation stress is related to the longshore current, the axes will become parallel to the coastline.)

The radiation stresses were derived from the linear wave theory equations by integrating the dynamic pressure over the total depth under a wave and over a wave period, and subtracting from this the integral static pressure below the still-water depth for unit width:

$$\int_0^{d+\eta} (p + \rho V^2) dz \tag{3.2}$$

where d: water depth, η: instantaneous water level, p: pressure, ρ: water density and V: velocity

This integral is equivalent to the X-component of the force acting in this plane; its unit, in the SI-system, is kgm/s^2, or N (Newton). The wave-induced contribution to the time average value of the integral per unit width is, by definition, the XX-component of the radiation stress, written as S_{XX}:

$$S_{XX} = \int_0^{d+\eta} (p + \rho V^2) dz - \int_0^d P_0 dz \tag{3.3}$$

in which the hydrostatic contribution is equal to:

$$\int_0^d P_0 dz = \int_0^d \rho g z \, dz = (1/2)\rho g d^2 \tag{3.4}$$

In the notation S_{XX}, one subscript (X) stands for the direction of transfer (through a plane X = constant) and the other for the component of momentum being transferred (X) (Fig. 3.6).

The value of S_{XX}, defined in Eq. (3.5), can be calculated from any wave theory. When we carry out this integration — it is a considerable task — over the depth on a plane perpendicular to the X-axis, the result is then

$$S_{XX} = \left[\frac{2kd}{\sinh 2kd} + \frac{1}{2}\right] E \qquad (3.5)$$

where, S_{XX}: the principal radiation stress component in the direction of wave propagation, k: wave number $(2\pi/L)$, L: wave length, E: wave energy. S_{XX} can therefore be denoted by:

$$S_{XX} = (2n - 1/2)E \qquad (3.6)$$

Note: Although S_{XX} is proportional to the wave energy E, it does not imply as energy per unit area. Its physical meaning is that of a rate of momentum transfer per unit width, or a force per unit width (SI-unit: N/m). Equation (3.6) is adopted for practical applications.

Computation of the second principal radiation stress component acting on a vertical plane perpendicular to the wave yields:

$$S_{YY} = \frac{kd}{\sinh 2kd} E \qquad (3.7)$$

or (for sinusoidal, progressive free gravity surface waves) expressed in terms of n,

$$S_{YY} = (n - 1/2)E \qquad (3.8)$$

Application of the usual approximations for deep water ($n = 1/2$) yields:

$$S_{XX} = (1/2)E, \qquad S_{YY} = 0 \qquad (3.9)$$

In shallow water ($n = 1$) these stresses become:

$$S_{XX} = (3/2)E, \qquad S_{YY} = (1/2)E \qquad (3.10)$$

The most important parameter influencing the radiation stress is the wave height. In deep water, this is the only influencing factor. In intermediate wave height, the water depth, d, and wave length, L, (via k), or simply "n" are also important. In shallow waters, it appears that the radiation stress depends only on the wave energy, in turn the wave energy is a function of water depth including at the breaker zone. If we now consider a square column of water enclosed by four vertical principal planes shown in Fig. 3.6 then, if the wave conditions and depth at all four planes 1, 2, 3, 4 are identical, the radiation stress component on opposite sides of the "block" shown in the figure are identical and there is no resulting force.

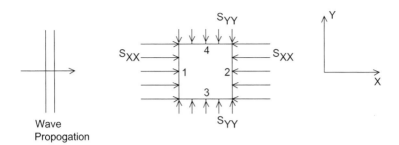

Fig. 3.6 Principal radiation stresses.

Only if the wave conditions vary between planes 1 and 2 or 3 and 4, there will be a resultant force. Thus, we can expect the radiation stress to influence physical processes only in areas where wave conditions change. Such areas would, therefore, be at locations where wave refraction, diffraction, shoaling, or breaking occurs.

The following example illustrates these changes in the principal radiation stresses caused, in this case, only by shoaling and breaking of the waves as they approach the shore. The waves approach the shore with wave crests parallel to the shoreline.

Note: The axes, used in radiation stress theories are different from the ones used in coastal engineering (see Fig. 3.3). One should be aware when and which particular axes are being used. While, the evaluation of H, S_{XX} and S_{YY} are presented in Problem 3.1, the variations of the said parameters along the water depth are projected in Fig. 3.7. Notice below the stresses "grow" as the waves approach the coast outside the breaker zone, and how breaking reverses this growth process.

If the waves approach the coast at an angle α, the principal radiation stresses, S_{XX}, acts in the direction of wave propagation and S_{YY} perpendicular to this direction. Since the coastal processes to be studied in later chapters of this syllabus can be split into components parallel and perpendicular to the coastline, it is convenient to work with radiation stress components along these axes. Figure 3.8 shows a plan view of a coastal area with principal stresses acting on an element oriented parallel to the wave crests and normal and shear stresses on an element parallel to the coastline.

Figure 3.9 shows the Mohr Circle for S_{xx} and S_{yy} as shown in Fig. 3.7. Those familiar with the pole method of using the Mohr Circle will recognize that the pole is at S_{XX} and the stresses on a plane located at an angle θ

can be found by passing a line having the same relative orientation through the pole. The mathematical description can more easily be obtained either from a force equilibrium on the element in radiation stresses for oblique approaching waves or from the geometry of the circle.

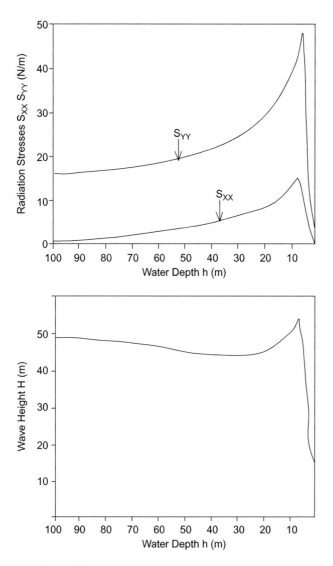

Fig. 3.7 S_{XX}, S_{YY} and wave height as a function of water depth.

3. Sediment Transport

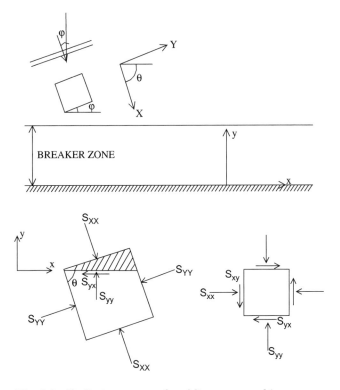

Fig. 3.8 Radiation stresses for oblique approaching waves.

In either case, the results are:

$$S_{xx} = \frac{S_{XX} + S_{YY}}{2} + \frac{S_{XX} - S_{YY}}{2} \cos 2\theta \qquad (3.11)$$

$$S_{yy} = \frac{S_{XX} + S_{YY}}{2} - \frac{S_{XX} - S_{YY}}{2} \cos 2\theta \qquad (3.12)$$

$$S_{xy} = \frac{S_{XX} - S_{YY}}{2} \sin 2\theta = \frac{S_{XX} - S_{YY}}{2} \sin 2\varphi = S_{yx} \qquad (3.13)$$

wherein, S_{xx}: radiation stress acting on a plane perpendicular to the coast, S_{yy}: radiation stress acting on a plane parallel to the coast, S_{xy}: radiation shear stress acting on a plane perpendicular to the coast, S_{yx}: radiation shear stress acting on a plane parallel to the coast, θ: angle between the two axis systems, φ: angle between wave crest and coast line.

The components S_{xx} and S_{yy} are similar to a normal stress, and S_{xy} and S_{yx} to a tangential stress, or shear stress.

A model developed by Losada et al. [1982] balances driving and resistance terms (gradients of radiation and turbulent Reynolds stresses and bottom friction) to get a general expression for the velocity. This equation shows explicitly the influence of Iribarren's parameter on longshore current generation. It has been tested with field and laboratory data, obtaining a reasonable fit to measured values. The resulting longshore momentum equation by Mei (1983) is given as,

$$\frac{\partial S_{xy}}{\partial x} + \frac{\partial S_{xy}^t}{\partial x} + R_{yb} = 0 \qquad (3.14a)$$

in which, S_{xy} = Excess momentum flux due to wave oscillation, S_{xy}^t = Excess momentum flux due to turbulent fluctuations and R_{yb} = y-component of the horizontal shear stress on the bottom.

The formulation of general current formula (V) is simply based on equilibrium between radiation stress and bottom friction. The velocity distribution in the breaker zone can be achieved with friction stress as a function of distance from the coast.

$$V = \frac{5\pi}{8\sqrt{2}} \frac{\sin(\alpha)}{C_0} \frac{C}{\sqrt{f_w}} \gamma \left(\sqrt{g}\right) dm \qquad (3.14b)$$

in which, $\frac{5\pi}{8\sqrt{2}} = 1.388$, $\frac{\sin(\alpha)}{C_0}$ depends only on the deep wave condition, $\frac{C}{\sqrt{f_w}}$ is a friction term dependent upon the bottom roughness, water depth and the local wave condition, γ depends on the wave condition and beach slope, m.

3.7 Cross-shore Sediment Transport

In general, any beach will be in equilibrium where the net sediment transport is zero under constant wave conditions. This equilibrium beach slope will increase with increasing grain size. In contrast, for a given grain size, an increase in the wave steepness will decrease the equilibrium beach slope. Normally, when the waves break, sediments materials are thrown into suspension and transported up to the beach in the direction of the movement of flow due to waves or currents or a combination of both. Also the gradation of the beach materials are usually with huge materials on the highest part of beach and grain size reduces seaward. During storm activities, waves will be high with steep-fronted of high frequency. Such a case will result in rapid and drastic removal of sea bed materials for a short term. A detailed discussion of sediment transport during storm surge is given in Sec. 3.10.

The shape of the waves influences the cross shore sediment transport. In shallow water, waves are mostly in asymmetrical form with crest velocity is more which are directed towards onshore and trough velocity is lower, directed towards offshore. These flows results in suspended sediments to move shoreward and bed load sediments seaward. As waves propagate form offshore to shallow water it deforms, in addition it experiences transformation due to its breaking and thereafter, the dissipation of energy takes place rapidly, i.e., the wave enters the wave decay region. The region inshore of the breaker point is divided into a breaker transition and broken wave regions. Based on wave propagation, Larson and Kraus [1989] divided the shallow region of coast into four zones as shown in Fig. 3.9 showing different hydrodynamic properties and contribute to different influence and relationships for sediment transport.

The pre-breaking zone (I) is from the depth seaward of significant sediment transport to the breaker point. The breaker transition Zone (II) is from the breaker point to the point where waves plunges or curls. The zone of broken wave (III) is from the wave plunging point to the area of offshore end of the surf zone and the swash Zone (IV) is from the shoreward boundary of the surf zone to the shoreward limit of the run up.

The knowledge of both cross shore and along shore sediment transport is essential foundation for numerical modelling. Estimation of cross shore transport rates may be done through empirical relationships which for the different zones are,

Zone 1:

$$q = q_b \exp\{-\lambda_1(x - x_b)\} \quad (3.15)$$

Zone 2:

$$q = q_p \exp\{-\lambda_2(x - x_p)\} \quad (3.16)$$

q — net cross-shore transport rate (m^3/m/sec)
$\lambda_{1,2}$ — spatial decay coefficients in transport Zones 1 and 2, (m^{-1})
X — cross-shore co-ordinate from the seaward end of the beach profile

The subscripts b and p in Eqs. (3.15) and (3.16) denote quantities evaluated at the breaking point and the plunging point, respectively. Spatial decay coefficients λ_1 and λ_2 are estimated empirically [Larson and Kraus, 1989] in the sediment transport rate in the same above equations are given by,

$$\lambda_1 = 0.4(D_{50}/H_b)^{0.47} \quad (3.17)$$

$$\lambda_2 = 0.2\lambda_1 \quad (3.18)$$

where, D_{50} is the median grain size in mm and H_b is the breaking wave height in m.

Zone 3:

Based on the assumptions of Kriebel and Dean [1985] that the cross-shore transport rate is proportional to the excess energy dissipation per unit volume over a certain equilibrium value of energy dissipation (D_{eq}) in fully broken waves, which was defined by the amount of energy dissipation per unit volume that a beach with a specific grain size could withstand without generating significant sediment transport. In other words, the cross-shore transport rate "q" in fully broken wave region with horizontal sea floor, is expressed as,

$$q = K(D - D_{eq}) \tag{3.19a}$$

$$D = (1/h)(dF/dx) \tag{3.19b}$$

$$D_{eq} = (5/24)\rho g^{3/2} \Gamma^2 A^{3/2} \tag{3.19c}$$

$$h = d + \eta \tag{3.19d}$$

d: water depth (m) from the undisturbed free surface, η: set up/set down (m), A is the Brunn's shape parameter, mainly a function of D_{50} and K: transport coefficient (m^4/N).

As the transport rate also depends on the local slope of the sea floor, an extra term is added to account for the effect of the local slope. With this modification the transport rate becomes,

$$q = K[D - D_{eq} + (\varepsilon/K)(dd/dx)], \quad D > D_{eq} - (\varepsilon/K)(dd/dx) = 0,$$
$$D < D_{eq} - (\varepsilon/K)(dd/dx) \tag{3.20}$$

where "ε" is the slope related transport rate coefficient (m^2/sec).

Zone 4:

The transport rate in the swash zone is assumed to decrease linearly from the end of the surf zone (Zone 3) to the run-up limit given by,

$$q = q_z\{(x - x_r)/(x_z - x_r)\} \tag{3.21}$$

where subscripts "z" and "r" are for quantities evaluated at the end of the surf zone and run-up limit, respectively. The active sub-aerial profile height "z_r" or z-coordinate of the run-up limit of waves is determined using an empirical relation derived from Large Wave Tank (LWT) experiments [Kraus and Larson, 1988] given by,

$$Z_r/H_0 = 1.47\, \xi^{0.79} \qquad (3.22)$$

where ξ is the surf similarity parameter $= \tan \beta/(H_0/L_0)^{1/2}$.
The slope to be adopted in this formula is the average beach slope. The corresponding x-coordinate "x_r" of the run-up limit in Eq. (3.21) is interpolated from the profile depths defined at grid points.

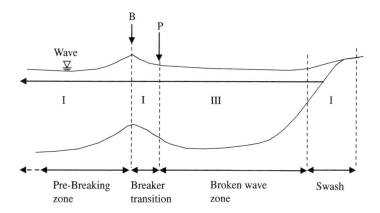

Fig. 3.9 Sediment transport zones.

3.8 Longshore Sediment Transport

3.8.1 Nearshore currents responsible for sediment transport

The wave-induced current systems are generally recognized in the near shore zone, which, dominate the water movements in addition to the to-and-fro motions produced by the wave orbits directly. They are (a) a Cell Circulation system of rip currents and feeding longshore currents and (b) longshore currents produced by an oblique wave approach to the shoreline. These are illustrated in Fig. 3.10. The longshore currents are mainly responsible for the transport of sediments along the shore and it is called longshore sediment transport.

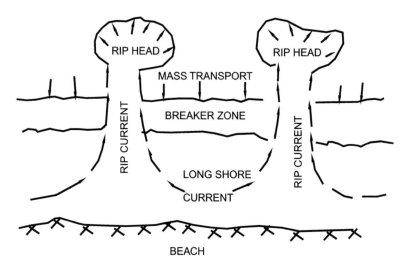

Fig. 3.10   Near shore current system.

3.8.2   *Phenomena of littoral drift*

A wave from Deep Ocean characterised by its height, H and length, L when moves towards the shore, its length initially undergoes a reduction till the wave reaches a depth of nearly 0.16 times its length. Thereafter, the wave height starts increasing until the water depth is about 1 to 1.5 times the water height. The cross-section of sediment transport mechanism is brought out in Fig. 3.11.

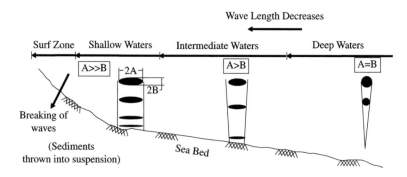

Fig. 3.11   Mechanics of sediment transport.

The orbital paths of water particles which are nearly circular till the depth is half the wave length become elliptical shorewards of this point. Shoreward, there will be water particle movements at the bottom and the wave is said to feel the bottom. The velocity and acceleration of particle movement increases with decreasing depths and at a depth depending on the sediment size and wave characteristics, the bottom sediments are put into motion. At this point, the material movement will be relatively small; but with decreasing depth, the movement increases and near the breaker, a considerable quantity of sediments are thrown into suspension due to the high turbulence associated with the breaking of waves. These are then easily moved by even slow moving currents.

This movement of material both in suspension and as bed load alters the existing profile. The change goes on till equilibrium is reached between the wave system which is being altered by the temporally changing depths and the bed profile which is similarly changed by the waves. However, equilibrium is seldom reached because the waves are continuously changed by meteorological factors. Sediment transport mechanism in plan view is depicted in Fig. 3.12.

Considering the action of waves on any beach in plan, the wave crests reaching the shore are seldom parallel to the shoreline or the underwater contours. The effect of this oblique attack of the waves on the shore is to generate two components of the fluid velocity, of which, one along the direction parallel to the shore is called longshore currents responsible for the transport of sediment along the shore. This is referred to as "longshore sediment transport".

The second component of the velocity in the direction normal to the shore, transport the sediment in the direction perpendicular to the shoreline. This mode of sediment transport is referred to as "onshore-offshore sediment transport" or cross-shore transport. The longshore sediment transport, however, is more dominant and mainly responsible for the shoreline instabilities.

The sediments especially the material thrown into suspension at the breaker zone is easily transported by the longshore currents. Apart from this, the zig-zag path described by the water mass and the sediments in the foreshore due to the uprush and backrush from the breakers also cause transport of material along the shore. The transport of material in the longshore direction by waves and currents near the shore is known as littoral drift, sometimes the material so transported is also called by the same name.

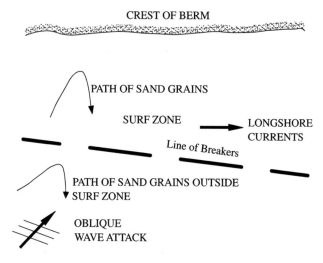

Fig. 3.12   Mechanics of sediment transport (plan view).

3.8.3   *Estimating longshore transport*

Unfortunately, quantitative estimation of sediment transport rates is extremely difficult. Short term or long term changes in the beach width or any changes in volume of materials can be easily identified by field survey or even aerial surveys. Long term study reveals trend of accretion or depletion in a particular location. However, these results based on trends are not certainly a direct estimation of longshore transport rate along the coast. Whereas, it dictates the imbalance in the coast and also the net direction drift of sediments. Marine structures constructed perpendicular to the shore dictates the sediment supply and significant changes in the beach. Direct measurement of longshore transport has been attempted using a variety of techniques, such as deposition of a tracer material (radioactive, dyed or artificial sediment) or installation of traps.

3.8.4   *CERC method*

The empirical relationship between the longshore component of wave energy flux entering the surf zone and the immersed weight of sand moved with a non-dimensional coefficient "K" is termed as CERC formula. The key assumption behind this method is that in the surf zone the longshore transport rate, Q depends on the longshore component of energy flux. Waves approaching to shore at an angle (α) are the primary cause for longshore sediment transport. Waves approaching perpendicular to the shore responsible for cross-shore sediment transport and oblique wave responsible for longshore sediment transport are brought out in Fig. 3.13.

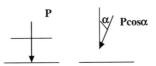

Fig. 3.13 Wave direction normal to the shore and at an angle (α) to the shore.

Where P is the wave power, given as,

$$P = EC_g = (1/8)\rho g H^2 C_g \qquad (3.23)$$

As the alongshore wave power component breaks into two components, it can be re written as

$$\text{Alongshore} = P\cos\alpha \cdot \cos(90-\alpha) = P\cos\alpha \cdot \sin\alpha$$

Hence, the wave crests make an angle, α with the shoreline, the energy flux becomes,

$$P\cos\alpha = \frac{\rho g}{8} H^2 C_g \cos\alpha \qquad (3.24)$$

α = angle between wave direction and shore normal and the longshore component is given by

$$P_{ls} = P\cos\alpha\sin\alpha = \frac{\rho g}{8} H^2 C_g \cos\alpha\sin\alpha, \qquad P_{ls} = \frac{\rho g}{16} H^2 C_g \sin 2\alpha,$$

(since, $\sin 2\alpha = 2 \cdot \cos\alpha \cdot \sin\alpha$)

$$(3.25)$$

Based on the approximation at breaker line, the equation can be written as,

$$P_{ls} = \frac{\rho g}{16} H_b^2 C_b \sin 2\alpha_b \qquad (3.26)$$

The above equation is valid only if there is a single wave train with one period and one height. However, most ocean wave conditions are characterized by a variety of heights with a distribution usually described by a Rayleigh distribution. For a Rayleigh distribution, the correct height to use in above equation is the root-mean-square height. Whereas, most wave data are available as significant heights, and therefore significant wave height is substituted in the equation by certain approximation.

Based on the assumption, longshore energy flux in the surf zone is approximated by conservation of energy and evaluating the energy flux relation at the breaker line. Further, by assuming that the waves follow Rayleigh distribution, P_{ls} is written as,

$$P_{ls} = \frac{\rho g}{16} H_{sb}^2 C_{gb} 2\alpha_b \qquad (3.27)$$

where, H_{sb} — significant wave height at breaker line.

The values of P_{ls} calculated using significant wave height at breaker line is nearly twice the value of the energy flux for sinusoidal wave heights described by Rayleigh distribution.

The longshore energy flux in the surf zone is obtained on the assumption that the energy flux is conserved followed by the evaluation of the energy flux relation at the breaker zone.

$$I_l = K P_{ls} \qquad (3.28)$$

where, I_l — immersed weight transport rate (force/time), K — dimensionless coefficient, and P_{ls} — the longshore energy flux (force/time).

In the present study the relation for breaker wave height given by Sunamura and Horikawa [1974] is considered, since it includes the seabed slope as a parameter

$$H_b = H_0 \times m^{0.2} \times \left(\frac{H_0}{L_0}\right)^{-0.25} \qquad (3.29)$$

C_{gb} — wave group celerity at the breaker line, in shallow waters $C_g = C$, given by

$$C_{gb} = C_b = \sqrt{g d_b}$$

α_b — wave approach angle w.r.t shore normal at the breaker line.

The immersed weight transport rate I_l is given by

$$I_l = (\rho_s - \rho) g a' Q \qquad (3.30)$$

where ρ_s — mass density of sand; a' — volume of solids/total volume, assumed as 0.6 for beach sands, Q — volume of sediment transport (m^3/day or month or year).

3. Sediment Transport

By substituting for I_l in CERC equation, we can get

$$Q = \frac{K}{(\rho_s - \rho)ga'}P_{ls} \qquad (3.31)$$

In the above equation, if "K" taken as 0.39 or 0.4 — Then it is CERC method and we need to use H_s, whereas, if "K" taken as 0.77 — Then it is KOMAR method, were we need to use H_{rms}. The estimation of sediment transport in the above formula depends on wave height and direction only. Field measurements of Q and P_{ls} are plotted in Fig. 3.14. Figure 3.15 gives lines of constant Q based on Eq. (3.31). [For details refer US Army Corps of Engineers, CEM 2002.]

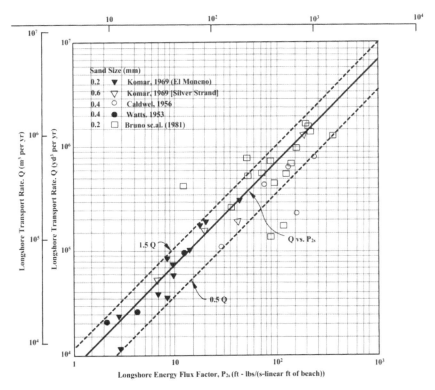

Fig. 3.14 Design curve for longshore transport rate energy flux factor (only field data included).

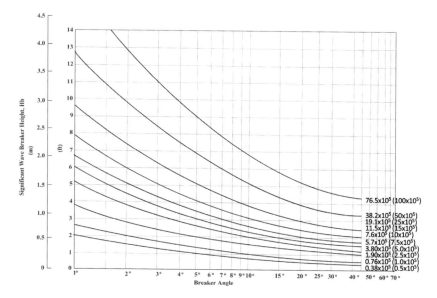

Fig. 3.15 Longshore sediment transport rate as a function of breaker height and breaker angle.

3.8.5 Calculation of P_{ls} using LEO data

LEO — Littoral Environment Observation field data collection program. It is another method for establishing the energy flux parameter. It is based on the visual observations of nearshore wave heights, wave periods and longshore current velocities. The detailed program is discussed by Berg [1969] and Schneider [1981]. Use of LEO data permits replacing the hard-to-measure wave angle term in equation with LEO longshore current measurements. The estimation of current is based on the time taken by the dye patch to travel a fixed distance in the surf zone.

The equations and example problem, which follow, are taken from Walton [1980], which presents derivations and additional references. The equation giving the longshore energy flux factor with LEO data variables is

$$P_{ls} = \frac{\rho g H_b W V_{\text{LEO}} C_f}{\left(\frac{5\pi}{2}\right)\left(\frac{V}{V_0}\right)_{LH}} \qquad (3.32)$$

where

$$\left(\frac{V}{V_0}\right)_{LH} = 0.2\left(\frac{X}{W}\right) - 0.714\left(\frac{X}{W}\right)\ln\left(\frac{X}{W}\right) \qquad (3.33)$$

3. Sediment Transport

and H_b = breaking wave height, W = width of surf zone, V_{LEO} = average longshore current due to breaking waves, C_f = friction factor (assume 0.01), X = distance to dye patch from shoreline. $(V/V_0)_{LH}$ is the dimensionless longshore current based on Longuet-Higgins [1970]. It is assumed that the LEO breaking wave height is a good approximation of the significant breaking wave height and that the mixing parameter in Longuet-Higgins' theory is 0.4.

The observation and field measured data based on LEO near Cochin Port, India is tabulated in Table 3.1.

Table 3.1. Typical measured and observed LEO data at Cochin Port, India.

UTM coordinates				Speed over ground (cm/s)	Course over ground (°)	Breaker angle (°)	Wave height (m)	Wave period (s)	Surf zone width (m)
Start point		End point							
Easting	Northing	Easting	Northing						
638478	1091273	638486	1091231	7.13	169	92	1	10	10 to 15
635607	1101679	635655	1101800	43.39	22	85	1.5	10	15 to 20
633396	1106072	633431	1105954	20.51	163	90	0.5	8	10 to 15
632066	1110943	632084	1110874	23.77	165	95	0.5	10	10 to 15

3.8.6 Method of Kamphuis [1991]

Kamphuis [1991] developed a wave transition model based on 170 sets of hydraulic model tests. Up to the breaking zone, wave transformation could be described by linear shoaling, refraction and bottom friction and in the braking zone using excess energy approach. These were verified with experimental and field results. There were several inconsistencies. It was stated that the differences in behaviour cannot be explained by common parameters such as wave steepness and surf similarity parameter and as such there is a need to look in to nonlinear shoaling effects. Based on dimensional analysis and calibration using laboratory and field data, the longshore transport as immersed mass (in kg/s) is given by:

$$Q_{t,im} = 2.33(T_p)^{1.5}(\tan\beta)^{0.75}(d_{50})^{-0.25}(H_{s,br})^2[\sin(2\alpha_b)]^{0.6} \quad (3.34)$$

$Q_{t,im}$ = longshore sediment (immersed mass) transport (kg/s), the dry mass is related to the immersed mass by $Q_{t,mass} = \rho_s/(\rho_s - \rho)$, $Q_{t,immersed\ mass}$; the conversion factor is about 1.64; $H_{s,br}$ = significant wave height at breaker line (m); α_b = wave angle at breaker line (°); $\tan\beta$ = beach slope defined as the ratio of the water depth at the breaker line and the distance from the still water beach line to the breaker line.

The value 2.33 is a dimensional coefficient related to the SI system assuming salt water ($1030\,\text{kg/m}^3$). The predictions through the formula of Kamphuis [1991] were consistent for both spilling and plunger breakers over CERC method of sediment transport estimation.

3.8.7 Sediment distribution across the surf zone

Wave-induced longshore currents at the mid surf-position have been evaluated using the relation suggested by Komar [1975] according to which,

$$V = 2.7\, u_m \sin\alpha_b \cos\alpha_b \qquad (3.35)$$

where $u_m = 1/2\gamma\, C$, is the maximum horizontal velocity of the waves, γ is the wave breaking index, a constant whose value is 0.78, C is the wave celerity in shallow water, α_b is the breaker angle.

It should be noted that the above equation is independent of the beach slope, though the longshore current velocity formula recommended by CERC [1984], based on Longuet-Higgins [1970] is dependent on beach slope. However, Komar [1979] has tested the dependence of longshore current and stated it is not directly proportional to the beach slope as stated in the CERC formula.

This method has some more basis than other two method, that the longshore current is computed [Longuet-Higgins, 1970] by equating gradient in the radiation shear stress to bottom friction, with assumption that, the shallow-water wave theory is valid only till the breaker line where the depth d is equal to h_b, the mean longshore current, in the absence of horizontal mixing. In the present study the coast has been assumed to be without groins for computing sediment transport using this method.

$$V_0 = \left(\frac{5\pi}{16}\right)\left(\frac{1}{C_f}\right)(gd_b)^{0.5}\gamma\varepsilon^2 m \cdot \sin\alpha_b \qquad (3.36)$$

where, m — the bed slope; C_f — current friction factor (around 0.01; Longuet-Higgins, 1970); γ — H/d and $\varepsilon = 1/(1 + 0.375\gamma^2)$.

However in the real sea conditions, the propagating wave will be of random in nature and hence, there will be a lateral mixing due to different waves with varying period breaks consecutively. Hence, he proposed a solution as

$$V = \begin{cases} B_1 X^{p_1} + AX & \text{for } 0 < X < 1 \\ B_2 X^{p_2} & \text{for } 1 < X < \infty \end{cases} \qquad (3.37)$$

where X and V are in non-dimensional form.

3. Sediment Transport

$X = x/x_b$, x — distance normal to the shoreline, x_b being the distance from the shoreline to the breaker zone and V — proportionality coefficient obtained with the inclusion of lateral mixing, needs to be multiplied with V_0 to obtain the actual velocity.

The other parameters are given by

$$A = [1/(1 - 2.5P)]; (P \neq 0.4); \quad B_1 = [(P_2 - 1)/(P_1 - P_2)]A;$$

$$B_2 = [(P_1 - 1)/(P_1 - P_2)]A;$$

$$P_1 = (-3/4) + [(9/16) + (1/P\varepsilon)]^{1/2}; \quad P_2 = (-3/4) - [(9/16) + (1/P\varepsilon)]^{1/2}$$

$$\varepsilon = 1/(1 + 0.375\gamma^2); \quad P = (\pi m N/\gamma C_f)$$

where, γ — is the wave breaking index (consider as $H = 0.78d$).

As you see in the above equations all the constants depends on the nondimensional parameter "P", which again depends on the lateral mixing parameter "N" varies between 0 and 0.016. Komar [1979] has combined the above longshore current velocity distribution of Longuet-Higgins [1970] with Bagnold [1966] and formulated the distribution of sediment transport along the surf zone due to waves and longshore current as

$$I_i = K_2[C_f \rho V^2 + 0.5 f \rho (0.25 \gamma^2 g d)]V \tag{3.38}$$

where, apart from the variables explained above, f — coefficient for oscillatory wave motion, V — local longshore current velocity, K_2 — proportionality constant between available power and resulting sediment transport. This K_2 is evaluated by integrating the above equation across the surf zone gives the equation below,

$$I_l = K_2 \left(\frac{1}{8}\rho\gamma^2 fg\right) \left(\frac{A}{3} + \frac{B_1}{P_1 + 2}\right) V_0 X_b^2 \tan\alpha$$

$$+ K_2 C_f \rho \left(\frac{A^3}{4} + \frac{3A^2 B_1}{P_1 + 3} + \frac{3AB_1^2}{2(P_1 + 1)} + \frac{B_1^3}{(3P_1 + 1)}\right) V_0^3 X_b \tag{3.39}$$

where A, B_1 and P_1 are from longshore current distribution by Longuet-Higgins (1970), V_0 — being the longshore current, in the absence of horizontal mixing, X_b — surf zone width and this Eq. (3.78) will be equated to total transport rate given by Komar and solved for K_2.

$I_l = KP_{ls}$ ($K = 0.77$) gives rise to

$$KP_{ls} = K_2 \left(\frac{1}{8}\rho\gamma^2 fg\right)\left(\frac{A}{3} + \frac{B_1}{P_1+2}\right) V_0 X_b^2 \tan\alpha$$

$$+ K_2 C_f \rho \left(\frac{A^3}{4} + \frac{3A^2 B_1}{P_1+3} + \frac{3AB_1^2}{2(P_1+1)} + \frac{B_1^3}{(3P_1+1)}\right) V_0^3 X_b \quad (3.40)$$

Once K_2 is obtained, it will be substituted in Eq. (3.78) to get the sediment transport distribution along the surf zone.

3.8.8 Van Rijn method

Van Rijn [1993] studied the transport process of sand under combined action of currents and irregular waves experimentally in a flume. Fine and medium fine sand were used as sediment material. The current velocities for following and opposing waves were generated. It was found that average velocities in the near bed region decreases with waves following or opposing current. It had negligible effect on concentration and transport rate.

$$Q = 40\, K_{swell} K_{grain} K_{slope} (H_{sb})^3 \sin(2\alpha_b) \quad (3.41)$$

Q = alongshore sediment transport (kg/s), H_{sb} = significant wave height at breaker line (m), $\tan\beta$ = slope, α_b = breaker angle (deg), swell correction factor $K_{swell} = T_P/6$, K_{grain} = particle size correction factor = $0.20/D_{50}$, K_{slope} = slope correction factor = $(\tan\beta/0.01)^{0.5}$.

3.9 Sediment Cell Concept and Its Application

As the coastline is continuously changing under the influence of hydrodynamic forces such as the waves, currents, wind, tides and other physical forces that interact with the coastal area, it is highly dynamic in nature. The coastline adjoining a continent/island, even though being continuous cannot be considered as a single unit mainly because of the variations observed in coastal morphology and the associated coastal processes. The spatial and temporal variations in the coastline characteristics can be attributed to the spatial variation in coastal morphology, nearshore wave characteristics, movement of sediments etc. For addressing this issue, which is very important for planning/designing of coastal/shoreline management measures the coastal stretch can be conveniently divided into a series of inter-linked physical units of manageable lengths. Each of these individual coastal units represents a pre-defined length of coastal area consisting of both offshore and nearshore components. These individual units can be considered as inter-linked segments of the original coastal line and the assumption is that the sediment transport within each segment/unit is mostly self-contained with minimum influence on the adjacent/neighbouring cells. However each cell can have its own sources (inputs) and sinks (outputs). The division of the coast into manageable segments has to be done carefully considering the spatial and temporal variations in the coastal geomorphology as well as the associated nearshore sediment transport processes. This approach is known as the sediment cell concept. The sediment cell concept proposed by Van Rijn [1997, 2011] has been widely adopted for coastal engineering applications. Adopting this method the coastal stretch can be conveniently subdivided into smaller self-contained cells or compartments without losing/ sacrificing its integrity. It has been well established that the nearshore sediment movements occur in discrete areas of cells called sediment cells, within which inputs and outputs are balanced.

A sediment cell is defined as a stretch of coastline and its nearshore area within which the movement of coarse sediments is largely self-contained. In other words, these are distinct areas of coastline (including the offshore part) separated from the adjoining coastal stretches by well-defined boundaries, such as natural headlands, artificial structures projecting into the sea, natural inlets etc. The general assumption is that the movement of sediments within a cell will not have any major effect on the adjacent cells. Even though theoretically, sediment cells are considered as closed systems without any loss or gain, in reality, there is a possibility that some of the fine sediments may move around headlands or obstructions and eventually enter into the neighbouring cells. Hence, wherever there is a significant change in the coastal processes within a sediment cell, sub-cells within the major cells can be considered and further division of sub-cells also can be carried out if required at meso- and micro-level for a detailed investigation. This type of sub-cell level studies becomes necessary, when, the coastal processes differ significantly at certain locations within a cell. The cell and sub-cell boundaries are usually defined at locations, where, there is a discontinuity in the transport rate or change of direction (probably could be due to the presence of headlands, spits, river mouths etc.). These boundaries need not be always of natural origin. Hard structures like jetties, breakwaters, long groins etc. which are normally shore connected can be considered as artificial boundaries for defining a sediment cell especially when the structures are long enough to extend beyond the surf zone.

The concept of Sediment Cell is explained taking the Kerala coast located along the SW coast of India as an example. Adopting the sediment cell approach, the entire coastal stretch of Kerala which is approx. 590 km can be divided into 4 major sediment cells. The key parameters considered for the division are the spatial variation in nearshore bed slope, variation in shoreline orientation, nearshore wave characteristics, coastal morphology etc. the details of which are presented in Table 3.2. The table gives the shoreline orientations with respect to North, the nearshore wave climate, the average beach slope and the average sediment grain size of the four major sediment cells (SC) designated as SC-I to SC-IV as presented in Fig. 3.16. The southernmost cell SC-I can be considered as a high energy sector with the Poovar inlet in Trivandrum and Thangassery headland at Quilon as the southern and northern boundaries, respectively. For the second cell

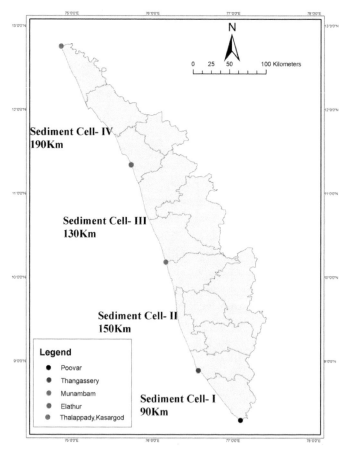

Fig. 3.16 Major sediment cells — Kerala coast.

SC-II, with medium to low energy conditions, the southern and northern boundaries are the Thangassery headland and Munambam inlet in Cochin, respectively. The third and fourth Sediment Cells — SC-III and SC-IV, which again, represent a combination of medium to low wave energy conditions have the Nandhi headland in Calicut and the Manjeswar inlet in Kasargod as their respective northern boundaries, whereas, the southern boundary of these cells are same as the northern boundary of the previous cell.

Table 3.2. Shoreline orientation, wave condition, nearshore bed slope and grain size of each sediment cell.

Sediment cell	Average shoreline orientation (° w.r.t N)	Nearshore wave climate	Average nearshore slope	Average grain size (mm)
I	228	High	0.005–0.05	0.34–0.37
II	255	Medium to low	0.005–0.003	0.24–0.31
III	250	Low, medium to low	0.003–0.004	0.20–0.30
IV	248	Low to medium	0.002–0.004	0.24–0.3

3.10 Sediment Transport During Extreme Events

3.10.1 *General*

A hazard is a situation that possesses a level of threat to life, health, property, or environment and hence a coastal hazard refers to incidents which possess the same effect in the coast which is an interface between the land and shoreline. The hazards in the coastal environment can either be manmade or natural. The risk evoked by a natural hazard is measured by the impact it leaves behind on the people, facilities, services and structures in a coastal community. There are many coastal hazards including storm surge resulting in immediate erosion hazard and coastal inundation, coastal recession (long term erosion hazard), dune instability and sand movement, geotechnical hazards (for example, rock fall and landslides), sea level rise due to climatic change, harmful algal bloom, oil spills and tsunami.

Of all the coastal hazards, three major disasters such as storm surge, oil spills and tsunami, are discussed in detail in this chapter.

An abnormal rise in the sea level witnessed during a storm event, that is much higher than the predicted astronomical tide at a particular location is known as a storm surge. It is a coastal phenomenon, the surge is caused primarily by a storm's winds pushing water onshore, which results in huge quantities of water thrust along the coast. Storm tide is the total observed seawater level during a storm, which is a combination of storm surge and normal high tide. The low-lying coastal areas are extensively flooded which is often referred to as storm water inundation. Severest impacts are resulted when storm surge occurrence coincides with the highest high tide. Figure 3.17 describes the storm surge during highest high tide and storm during the low tide. At this instance, the inundated storm tide can flood the coastal areas within a period of several hours, which could have been free from flooding otherwise. Inundation caused due to storm tides accelerates erosion of sand dunes. It poses critical threat to property and infrastructure adjacent to the shore, which are not normally exposed to sea water. Tropical cyclones that occur with a warning through formation of low pressure zones attribute to the most devastating phenomena in coastal areas across the globe. The surge resulted from a cyclone greatly impacts the coastlines, bays and estuaries. Owing to the excessive flooding in the low-lying areas, a huge threat to human life, property and ecosystem is resulted. Such events are more common along the northern Gulf of Mexico and Bay of Bengal.

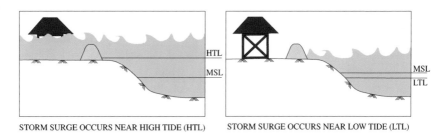

Fig. 3.17 Storm surge during different tide levels.

The extent of storm surge in a particular stretch of the coast depends by a number of parameters such as water surface elevations, wind intensity, central pressure, forward speed, storm size (radius of the maximum winds and gale force wind radii), angle of approach toward the coast, and the shape (sharpness, decay and extent) of the wind profile. The characteristic coastline features i.e. the geometries of bays, estuaries, channels in bathymetry etc. also influences the water level heights.

3. Sediment Transport   69

Indian National Centre for Ocean Information Services (INCOIS) has categorized the entire Indian coast into 4 zones

- Very high risk zones, VHRZ (Surge height >5 m)
- High risk zone, HRZ (Surge height between 3–5 m)
- Moderate risk zone, MRZ (Surge height between 1.5 m to 3 m)
- Minimal risk zone (Surge height <1.5 m).

The Indian Meteorological Department, classifies the tropical cyclones based on their wind speed intensity as follows

- Super cyclone storm >222 km/hr
- Very severe cyclone storm 118–221 km/hr
- Severe cyclone storm 88–117 km/hr
- Cyclone storm 62–87 km/hr
- Deep depression 52–61 km/hr
- Depression ≤51 km/hr

3.10.2 Case study

The Thane cyclone (29–30 December 2011) which hit the east coast of Indian Peninsula was one of the most devastating cyclone that resulted in significant damages along the Tamil Nadu coastline (13°26′40″N and 80°19′34″E to 8°17′41″N and 77°05′37″E). Waves as high as 8–12 m in a water depth of 20 m have been measured. Such huge waves, combined with a storm surge of 0.5 m, lead to severe damages to coastal structures during the passage of the cyclone.

A measurement campaign to measure the characteristics of waves and tides to understand the flow characteristics and sediment dynamics in between an existing groin field serving as coastal protection on the north of Chennai harbor (at 13°N) has captured the dynamics of a cyclone, Thane, that had encountered during 25–30 December 2011. The flow field, that is, orbital velocities normal (u) and parallel (v) to shoreline, and vertical (w) have been measured using a directional tide gauge in a shallower water depth of 5.0 m (St-4) and in a water depth of 8 m (St-2). A wave rider buoy, commissioned in a water depth of 20 m (St-1), had recorded the directional wave climate at an offshore distance of 6.5 km. The tide gauge is of pressure sensor type, from which the surface elevation has been derived. It was of bottom mounted type and hence, not suitable for deepwater deployment to measure the wave climate. Any such measurement in deeper water might result in the loss of information in capturing high frequency waves. A wave rider buoy on the other hand floats on the surface. The buoy heave motion provides the surface elevation and the directionality can be derived from its roll and pitch. Hence it has been deployed in a water depth of 20 m. A view of the instruments that were deployed in Ocean for the acquisition of field data is shown in Fig. 3.18. The sampling rate employed for the data collection was 2 Hz for duration of 30 minutes. Such measurements were done once in 3 hours at four locations. Figure 3.19 depicts the measurement locations on the bathymetry.

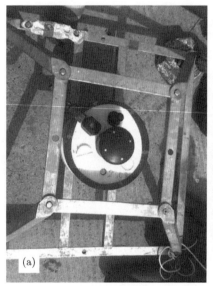

Fig. 3.18   Tide gauge (a) and wave rider buoy (b) during deployment.

A detailed statistical and spectral domain analysis of the time history of the wave surface elevation revealed the cyclone off the track wave growth along the Chennai coast. The integrated wave characteristics such as significant wave height (H_s), mean wave period (T_m) and mean wave direction (θ_m) have been derived from the directional spectra that were deduced from buoy and directional tide gauges. The variation of H_s, T_m and θ_m at different stations during 14–31 December 2011 is depicted in Fig. 3.20. It is seen that the H_s varies from about 1.5 m to 2.0 m, whereas, the T_m is found to be in the range of 6 s to 9 s. The said variations show a small degree

3. Sediment Transport 71

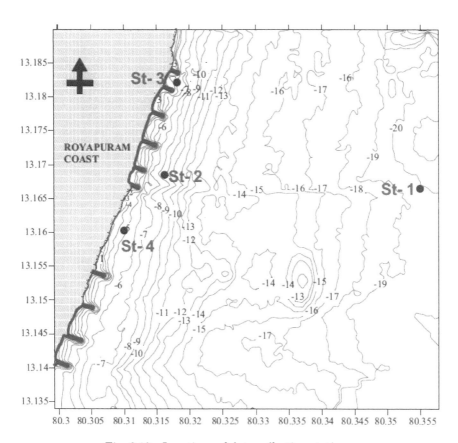

Fig. 3.19 Locations of data collection stations.

of variability. Subsequent drastic increase of H_s and T_m demonstrates the severity of the sea state during the approach of the Thane toward the coast. Its effect can be seen from the increase of H_s up to 6.2 m with an associated T_m up to 13 s in a water depth of 20 m.

Cyclone Thane, with the strength of a Category 1 hurricane of 65 knots (120 km per hour), struck south eastern India on December 30, 2011. Thane hit the coast of Tamil Nadu as shown in Fig. 3.21, and made landfall between Puducherry (11°56′N and 79°45′E) and Cuddalore (11°45′N and 79°45′E). Thane, commonly referenced as a cyclone instead of a hurricane or typhoon, formed in the Indian Ocean on December 25, 2011.

As seen in Fig. 3.22, cyclone Thane never formed a distinct eye, which usually becomes more apparent when the storm becomes a Category 2

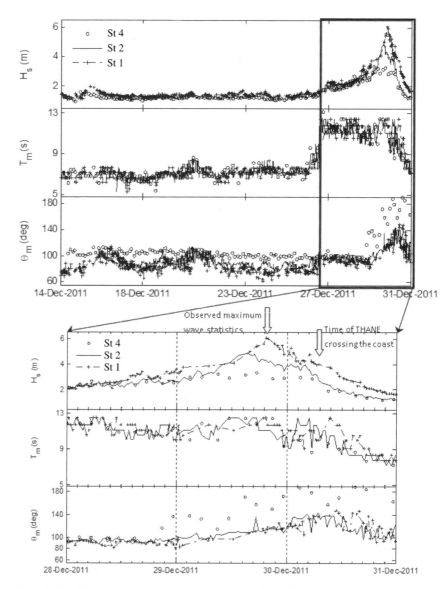

Fig. 3.20 Measured wave parameters during 14–31 December 2011 at the site.

storm with winds between 83–95 kt, or 154–177 km/hr. Satellite imagery shows well-defined spiral bands pushing on shore as the cyclone pushes to the west. Thane was downgraded into a tropical storm after it moved on shore.

3. Sediment Transport

Fig. 3.21 Satellite image of Thane cyclone.
(*Source*: Indian Meteorological Department)

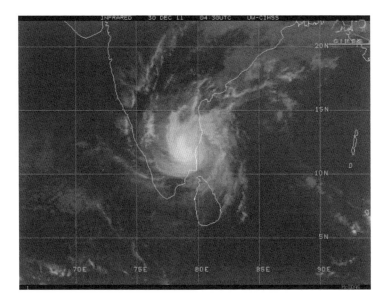

Fig. 3.22 Thane making landfall across southern India on 31 December 2011.
(*Source*: CIMSS)

3.10.3 Effect of sedimentation during extreme events

In the wake of an extreme event such as storm, cyclone or hurricane, the usual trend prevalent at a given site varies drastically. As a case study, the aftermath of Thane cyclone is discussed briefly.

Influence of Thane and Nilam cyclone characteristics on the wave climate

The influence in the wave climate due to change in the characteristics of cyclone is studied in detail. An in-depth analysis on this parameter has been carried out by considering two cyclones, Thane and Nilam, which has made the landfall near to the Chennai coast in the consecutive years 2011 and 2012, respectively.

Cyclone "Thane"

It has already been discussed about the field measured and simulated wave characteristics of Thane cyclone in the earlier section. However, in this parametric study, the directional aspects are compared for the two cyclones. The cyclone Thane can be considered as the strongest tropical cyclone over the past one decade. The cyclone had started gaining momentum from 26 December 2011. When it was centred at 9.5°N, 87.5°E, it had obtained further momentum by moving towards north-westwards. During its strongest state, it had centred at 12°N, 80.6°E during 1800 hours on 29 December 2011, with a maximum sustainable wind speed of 150 km/hr, prior to crossing the coast at Puducherry (11.8°N, 79.9°E) at 0730 hours on 30 December 2011.

Cyclone "Nilam"

The Nilam cyclone had started as a depression from the forenoon of 28 October 2012 when it was centred at 9.5°N, 86.0°E. It obtained momentum by moving towards western and north-western direction. It was at its strongest state when centred around 11.0°N, 81.0°E, during the forenoon of 31 October 2012, with a maximum sustainable wind speed of around 80 km/hr, prior to crossing the east coast of India near Mahabalipuram (12°37'36.84" and 80°11'33.72"E) at around 1630 hours on the same day.

The location of eye of the cyclone, central pressure drop and maximum sustained wind during the Thane and Nilam cyclones were obtained from IMD which are presented in Tables 3.3 and 3.4, respectively.

3. Sediment Transport 75

Table 3.3. Location, central pressure drop and maximum sustained surface wind velocity during the track of cyclone Thane (IMD).

Date	Time (UTC)	Eye centre Lat °N	Eye centre Long °E	Estimated central pressure (hPa)	Estimated max sustained surface wind (Knots)
26.12.2011	0000	9.5	87.5	998	30
	0600	10.0	87.5	998	30
	1200	10.5	87.5	998	30
	1800	11.0	87.5	996	35
27.12.2011	0	11.5	87.5	994	40
	0600	12.0	87.0	994	40
	1200	12.5	86.5	992	40
	1800	12.5	86.0	990	45

Table 3.4. Location, central pressure drop and maximum sustained surface wind velocity during the track of cyclone Nilam (IMD).

Date	Time (UTC)	Eye centre Lat °N	Eye centre Long °E	Estimated central pressure (hPa)	Estimated max sustained surface wind (Knots)
28.10.2012	0600	9.5	86.0	1004	25
	1200	9.5	85.0	1003	25
	1800	9.5	84.5	1002	25
29.10.2012	0000	9.5	84.0	1000	30
	0300	9.5	83.5	1000	30
	0600	9.0	83.0	1000	30
	1200	9.0	82.5	1000	30
	1800	9.0	82.0	1000	30
02.11.2012	0000	The system weakened into a well-marked low pressure area over Rayalaseema and neighborhood.			

The sediment transport rates were evaluated using two empirical methods namely CERC [1984], based on wave energy and Komar distribution method [1979], adopting theoretical distribution of alongshore current velocity across the surf width. From the recording stations deployed, as mentioned above, the breaker wave angle, breaker wave height and surf zone width have been derived with which, the breaker wave characteristics, the alongshore current velocity distribution followed by alongshore sediment distribution at three hourly interval are obtained. The monthly variations of alongshore current velocity and alongshore sediment distribution for November and December are shown in Figs. 3.23(a) and 3.23(b), respectively.

The sediment transport across the surf width for every 3 hours is summed up and plotted along with the sediment transport arrived from CERC formula in Fig. 3.24(a) and the same has been projected during

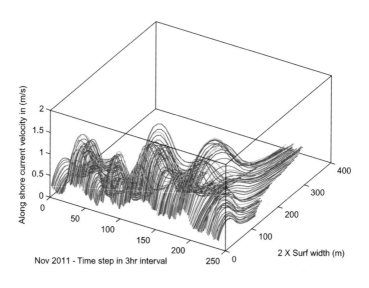

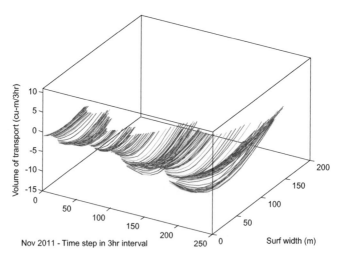

Fig. 3.23(a) Distribution of alongshore current velocity and sediment transport across surf zone for November.

3. Sediment Transport   77

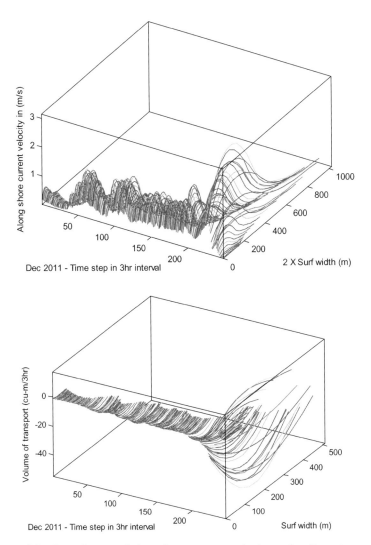

Fig. 3.23(b)   Distribution of alongshore current velocity and sediment transport across surf zone for December.

January to November for clear visualization in Fig. 3.24(b). The distribution for the month of December has been plotted separately in Fig. 3.24(c). The sediment transport across the surf zone is about 4,500 m³/3 hrs during normal wave climate and 45,000 m³/3 hrs during cyclone Thane. The foregoing results clearly bring out the effect of extreme event on the sediment transport in the coastal zone.

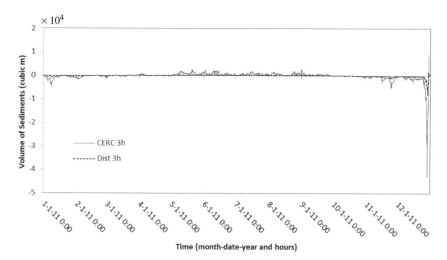

Fig. 3.24(a) Annual longshore sediment transport obtained from CERC and Komar distribution method (DIST), for the year 2011.

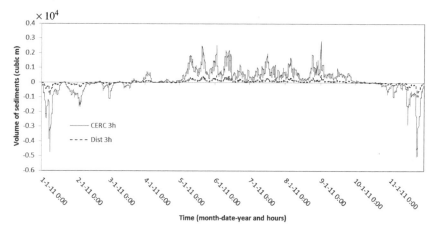

Fig. 3.24(b) Longshore sediment transport obtained from CERC and Komar distribution method (DIST), from January 2011 to November 2011.

The transitory variation during an extreme event directly affects the net sediment transport quantity and direction. Apart from the wave characteristics, the orientation of the shore, wind speed and angle, location of eye of the cyclone also influences the sedimentation.

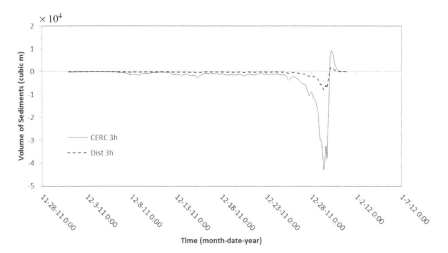

Fig. 3.24(c) Longshore sediment transport obtained from CERC and Komar distribution method (DIST) for December 2011.

Problem I

For the given data, $m = 0.01$ (bottom slope), $H_0 = 5$ m (deep water wave height), $T = 12$ s (wave period) calculate the S_{XX}, S_{YY}, H and p and show their variations along the depth.

Solution

$$L_0 = 1.56 \, T^2 = 225 \, \text{m} \quad \text{(deep water wave length)}$$

$$\xi_0 = \frac{m}{\sqrt{H_0/L_0}} = 0.067 \quad \longrightarrow \quad \text{spilling breakers}$$

$$L = \xi_0 0.17 + 0.08 = 0.71 \quad \text{(breaker index, Battjes [1974])}$$

Sample calculations for typical water depths are in Table 3.5, whereas, Fig. 3.7 shows the variations of the parameters over the water depth.

$$n = \frac{kd}{\sinh 2kd} + \frac{1}{2}, \quad E = (1/8)\rho g H^2$$

$$S_{XX} = (2n - 1/2)E$$

$$S_{YY} = (n - 1/2)E$$

$$\text{Pressure} = (1/2)\rho g d^2$$

Table 3.5. Radiation stress values.

Y (m) (1)	d (m) (2)	d/L_0 (3)	d/L (4)	kd (5)	L (m) (6)	n (7)	H/H_0 (8)	H (m) (9)	0.142 $\tanh(kd)$ (10)	E (N/m)	S_{xx} (N/m)	S_{yy} (N/m)	Pressure (N/m)
5000	50	0.321	0.331	2.08	151.2	0.565	0.9955	5.97	0.1376	44838	28264	2923	12507750
3000	30	0.192	0.219	1.37	137.2	0.677	0.9166	5.50	0.1249	38013	32447	6721	4502790
800	8	0.051	0.095	0.60	83.8	0.897	1.0182	6.11	0.0763	46906	60744	18645	320198.4
700	7	0.045	0.089	0.56	78.9	0.910	1.0425	5.70	0.0718	40808	53845	16720	245151.9
600	6	0.038	0.082	0.51	73.6	0.922	1.0722	4.88	0.0670	29982	40305	12657	180111.6
500	5	0.032	0.074	0.46	67.6	0.935	1.1106	4.07	0.0616	20821	28513	9052	125077.5
400	4	0.026	0.066	0.41	60.9	0.947	1.1622	3.26	0.0555	13325	18588	5963	80049.6

Problem II

Calculate the value of P_{ls} parameter for the following wave conditions using any four formulae for breaking wave characteristics:

$$H_0 = 1 \text{ m}, \quad \alpha_0 = 39°, \quad m = 1/50 \text{ and } T = 10 \text{ s}.$$

$$m = 0.02$$

$$L_0 = 1.56 \, T^2 = 1.56 \times 10^2 = 156 \text{ m}$$

$$C_0 = L_0/T = 15.6 \text{ m/s}$$

Table 3.6. Breaking wave heights and other parameters from different formulae.

Sl. No	Author	Formulae	H_b (m)	d_b (m)	L_b (m)	C_g (m/s)	$\sin 2\alpha_b$	P_{ls} (J/m-s)
1	Komar and Gaughan [1972]	$0.56 H_0$ $(H_0/L_0)^{-0.2}$	1.537	1.970	43.4	4.34	0.3446	2220.36
2	Sunamara and Horikawa [1974]	$H_0 m^{0.2}$ $(H_0/L_0)^{-0.25}$	1.616	2.072	44.37	4.44	0.3524	2567.86
3	Ogawa and Shuto [1984]	$0.68 H_0 m^{0.09}$ $(H_0/L_0)^{-0.25}$	1.69	2.167	45.23	4.52	0.3585	2908.52
4	Smith and Kraus [1990]	$H_0(0.34 + 2.47m)$ $(H_0/L_0)^{-0.3+0.88m}$	1.621	2.078	44.5	4.45	0.3532	2595.48

Problem III

A breaking wave of 1.5 m height approaches a beach slope of 1:100, at an angle 6° to the beach. Calculate and compare the longshore current profile for wave approaching with 11 sec and 6 sec by assuming that the friction factor may be taken as 0.002 and the mixing parameter N as 0.011. Where

$$A = [1/(1 - 2.5P)](P \neq 0.4)$$

$$B_1 = [(P_2 - 1)/(P_1 - P_2)]A$$

$$B_2 = [(P_1 - 1)/(P_1 - P_2)]A$$

$$P = (\pi m N/\gamma C_f)$$

$$\varepsilon = 1/(1 + 0.375\gamma^2)$$

m = beach slope, γ = breaker index = H_b/d_b, N = dimensionless constant, C_f = current friction factor.

$$P_1 = (-3/4) + [(9/16) + (1/P\varepsilon)]^{1/2}$$
$$\varepsilon = 1/(1 + 0.375\gamma^2) \quad \text{and}$$
$$P_2 = (-3/4) - [(9/16) + (1/P\varepsilon)]^{1/2}$$

The parameter P is non-dimensional and represents the relative importance of horizontal mixing.

Solution

Solving for $T = 11$ sec
Firstly find the breaking point using the equation below:
Maximum breaker height may also be approximated using criteria

$$\gamma = (H_b/d_b) = b - a(H_b/gT^2)$$

where

$$a = 4.46 \, g \, (1 - e^{-19m}),$$
$$b = [1.56/(1 + e^{-19.5m})],$$
$$m = 1/100, T = 11 \, and \, H_b = 1.5$$

For the present problem substituting the variables we get

$$a = 7.57, \, b = 0.855$$
$$\gamma = H_b/d_b = 0.855 - 7.57 \times [1.5/(9.81 \times 11^2)] = 0.846$$
$$d_b = 1.5/0.846 = 1.77 \text{ m}$$

Hence

$d_b = 1.77$ at a distance, $1.77 \times 100 = 177.3$ m
$\varepsilon = 0.788, \, \varepsilon^2 = 0.621$ and $f_c = 0.002$
$v_0 = (5\pi/16 f_c) \, \gamma(gd_b)^{1/2} \, m \, \varepsilon^2 \sin \alpha_b$
$ = [(5 \times \pi \times 0.846)/(16 \times 0.002)] \, (0.621)(9.81 \times 1.77)^{1/2} \, (1/100) \sin 6$
$ = 1.135$ m/s

$$P = (\pi mN/\gamma f_c)$$
$$= [(\pi \times 0.011 \times 1)/(0.846 \times 0.002 \times 100)]$$
$$= 0.204$$

$$P_1 = (-3/4) + [(9/16) + (1/P\varepsilon)]^{1/2} = 1.85$$
$$P_2 = (-3/4) - [(9/16) + (1/P\varepsilon)]^{1/2} = -3.35$$
$$A = [1/(1 - 2.5P)] = 2.04$$
$$B_1 = [(P_2 - 1)/(P_1 - P_2)]A$$
$$= [-3.35 - 1]/(1.85 - (-3.35))] \times 2.04 = -1.706$$
$$B_2 = 0.333$$

$$\left.\begin{array}{l} v = B_1(X)P_1 + AX \\ v = B_2(X)P_2 \end{array}\right\} \begin{array}{l} 0 < X < 1 \\ 1 < X < \infty \end{array}$$

we know $x_b = 177.3$ and $v_0 = 1.135$ m/s.

$X = x/x_b$ (1)	Dist from Shore (m) (2) = (1) × x_b	v/v_0 (3)	Longshore velocity $V = \text{Col}(3) \times v_0$ m/sec
0.1	17.73	0.17990211	0.20418889
0.2	35.46	0.32112706	0.36447921
0.3	53.19	0.42806967	0.48585908
0.4	70.92	0.50282289	0.57070398
0.5	88.65	0.54676862	0.62058238
0.6	106.38	0.56093082	0.63665648
0.7	124.11	0.54611816	0.61984411
0.8	141.84	0.50299594	0.57090039
0.9	159.57	0.43212749	0.4904647
1	177.3	0.333	0.377955
1.1	195.03	0.24197959	0.27464684
1.2	212.76	0.18079527	0.20520263
1.3	230.49	0.13827192	0.15693863
1.4	248.22	0.10787364	0.12243658
1.5	265.95	0.08561276	0.09717049

Similarly, longshore current for $T = 6$ sec is calculated and the results are plotted for comparison in the figure below.

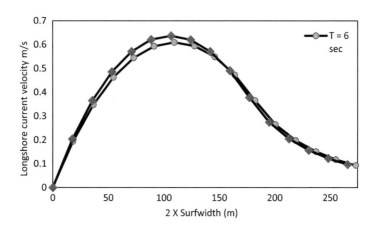

Problem IV

Find the net sediment transport for the data given below.

MONTH	Jan	Feb	Mar	Apr	May	Jun	Jul	Aug	Sep	Oct	Nov	Dec
H_0(m)	1.5	1.7	1.5	1.0	1.2	2.5	2.8	2.0	2.0	1.5	0.8	1.2
T	7	7	7	7	7	7	7	7	7	7	7	7
α_0	50	45	35	15	15	-45	-50	-30	-35	-45	20	20

Solution

- $P_{ls} = \rho g/16 \times H_b \times H_b \times C_b \times \sin(2\alpha_b)$, $H_b = H_0/3.33\,[(H_0/L_0)]^{1/3}$
- $d_b = H_b/0.78$, $C_b = L_b/T$, $\sin\alpha_0/\sin\alpha_b = C_0/C_b$
- Q (cu. m/month) $= 1290/12$ (cu.m-s/N-months) $\times P_{ls}$ (J/m-s)

Month	January	February	March	April	May	June	July
L_O	76.44	76.44	76.44	76.44	76.44	76.44	76.44
H_b	1.65	1.79	1.65	1.26	1.42	2.32	2.5
d_b	2.11	2.3	2.11	1.61	1.82	2.98	3.21
d_b/L_O	0.02765	0.03006	0.02765	0.02107	0.02381	0.03893	0.042
d_b/L_b	0.06832	0.07142	0.06832	0.05922	0.06313	0.08206	0.08553
L_b	30.93	32.18	30.93	27.19	28.83	36.26	37.53
C_b	4.42	4.6	4.42	3.88	4.12	5.18	5.36
$\sin \alpha_b$	0.31	0.3	0.23	0.09	0.1	−0.34	−0.38
α_β	18.05	17.45	13.29	5.16	5.74	−19.87	−22.33
$\sin 2\alpha_b$	0.63	0.57	0.45	0.18	0.2	−0.64	−0.7
P_{ls}	4729.85	5264.64	3378.46	690.11	1037.67	−11169.13	−14719.01
Q	508458.97	565948.78	363184.98	74186.33	111549.67	−200681.1	−1582293.3

(*Continued*)

(Continued)

Month	August	September	October	November	December
L_0	76.44	76.44	76.44	76.44	76.44
H_b	2	2	1.65	1.17	1.42
d_b	2.56	2.56	2.11	1.5	1.82
d_b/L_0	0.03352	0.03352	0.02765	0.01963	0.02381
d_b/L_b	0.0757	0.0757	0.06832	0.05706	0.06313
L_b	33.85	33.85	30.93	26.3	28.83
C_b	4.34	4.84	4.42	3.76	4.12
$\sin \alpha_b$	−0.22	−0.25	0.29	0.12	0.16
α_β	−12.7	−14.47	−16.85	6.89	9.2
$\sin 2\alpha_b$	−0.43	−0.48	−0.55	0.24	0.32
P_{ls}	−5194.62	−5798.65	−4129.23	772.71	1660.27
Q	−558422.18	−623354.99	−443892.75	83066.65	178479.48

Gross long shore sediment transport rate = 6293519.19
Net long shore sediment transport rate = −2523769.47
Negative sign shows that net transport of the long shore sediment is on south side.

Problem V

Based on LEO observations, the following values are assessed. Wave height, $H_{sb} = 1$ m, longshore current velocity, $V_{LEO} = 0.20$ m/s, width of surf zone, $W = 50$ m and distance of dye patch from the shoreline is, $X = 18$ m. Find longshore energy flux factor P_{ls}.

Solution

(a) Using Eq. (17) calculate $(V/V_0)_{LH}$

$$\left(\frac{V}{V_0}\right)_{LH} = 0.2\left(\frac{18}{50}\right) - 0.714\left(\frac{18}{50}\right)\ln\left(\frac{18}{50}\right) = 0.33$$

(b) Now, using Eq. (16), calculate P_{ls}

$$P_{ls} = \frac{(9.8)1025(1)(50)(0.20)(0.01)}{\left(\frac{5\pi}{2}\right)(0.33)} = 387.6 \text{ N/sec}$$

Longshore energy flux factor $= 387.6$ Jl/m-sec

Problem VI

With the data given, observe the influence of each parameter on P_{ls}:
(a) $H_0 = 1$ m, $\alpha_0 = 39°$, $m = 1{:}50$, $T = 10$ sec, (b) $H_0 = 1.5$ m, $\alpha_0 = 39°$, $m = 1{:}50$, $T = 10$ sec, (c) $H_0 = 1$ m, $\alpha_0 = 39°$, $m = 1{:}50$, $T = 15$ sec, (d) $H_0 = 1$ m, $\alpha_0 = 58.5°$, $m = 1{:}50$, $T = 10$ sec and (e) $H_0 = 1$ m, $\alpha_0 = 60°$, $m = 1{:}50$, $T = 10$ sec.
Calculate the longshore energy flux factor for all the above cases.

Solution

(a) $H_0 = 1$ m, $\alpha_0 = 39°$, $m = 1:50$, $T = 10$ sec, $L_0 = 1.56 \times T^2 = 1.56 \times 10^2 = 156$ m.

With formula of Sunamura [1983], refer to Table 3.6

$$H_b = 1 \times (1/50)^{\frac{1}{5}} \times (1/156)^{-\frac{1}{4}} = 1.616 \text{ m}$$

According to the condition of the breaking waves,

$$H_b = 0.78 d_b \Rightarrow d_b = \frac{H_b}{0.78} = \frac{1.616}{0.78} = 2.07 \text{ m}, \frac{d_b}{L_0} = \frac{2.07}{156} = 0.013$$

From the wave tables,

$$\frac{d_b}{L_b} = 0.04612 \Rightarrow L_b = \frac{d_b}{0.04612} = \frac{2.07}{0.04612} = 44.88$$

$$C_b = \frac{L_b}{T} = \frac{44.88}{10} = 4.488 \text{ m/sec},$$

$$C_0 = \frac{L_0}{T} = \frac{156}{10} = 15.6 \text{ m/sec}.$$

From Snell's Law,

$$\frac{\sin \alpha_0}{\sin \alpha_b} = \frac{C_0}{C_b}, \quad \frac{\sin 39°}{\sin \alpha_b} = \frac{15.6}{4.488}$$

$$\Rightarrow \alpha_b = 10°$$

Longshore energy flux factor is given by

$$P_{ls} = P_{lb} = \frac{1}{16}\gamma H_b^2 C_b \sin 2\alpha_b$$

$$= \frac{1}{16} \times (1025 \times 9.81) \times (1.616)^2 \times (4.488) \times \sin 2(10°) = 2519 \text{ N/sec}$$

Similarly,

(b) $H_0 = 1.5$ m, $\alpha_0 = 39°$, $m = 1{:}50$, $T = 10$ sec

Longshore energy flux factor = 6315 N/sec

(c) $H_0 = 1$ m, $\alpha_0 = 39°$, $m = 1{:}50$, $T = 15$ sec

Longshore energy flux factor = 3371 N/sec

(d) $H_0 = 1$ m, $\alpha_0 = 58.5°$, $m = 1{:}50$, $T = 10$ sec

Longshore energy flux factor = 3458 N/sec

(e) $H_0 = 1$ m, $\alpha_0 = 60°$, $m = 1{:}50$, $T = 10$ sec

Longshore energy flux factor = 3548 N/sec

From above, it is clear that increase in wave height increases P_{ls} value significantly. The increase in wave period and the angle also changes the value of P_{ls} parameter.

Problem VII

Evaluate P_{ls} (longshore energy flux component) using energy flux method, LEO (Littoral Environment Observation) method and compare the energy flux and sediment transport Q for the given data as below.

$\rho = 1024$ kg/m^3, $g = 9.81$ m/s^2, $C_f = 0.01$ (friction factor), $X = 50$ m (distance of dye patch from shoreline), Bed slope = 1:80, angle between shoreline geo-north $(u) = 650$. Therefore angle between shore normal and geo-north = 155°.

			Given data			$\vartheta°$
Month	H_0 (m)	T (s)	$\vartheta°$ w.r.t north		Measured V_{ieo} (m/s)	w.r.t shore normal
Jan	1.5	9	105		−0.32	−50
Feb	1.7	9	110		−0.31	−45
Mar	1.5	9	120		−0.28	−35
Apr	1.0	9	140		−0.09	−15
May	1.2	9	140		−0.14	−15
Jun	2.5	10	200		0.40	45
Jul	2.8	10	205		0.35	50
Aug	2.0	10	185		0.19	30
Sep	2.0	10	190		0.22	35
Oct	1.5	8	110		−0.38	−45
Nov	0.9	8	135		−0.12	−20
Dec	1.2	8	130		−0.18	−25

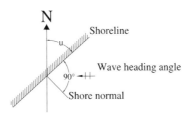

Solution

(a) Energy Flux Method

The detailed calculations for the month of January

$$P_{ls} = \frac{\rho g}{16} H_{sb}^2 C_{gb} \sin 2\alpha_b$$

where,

H_{sb} = wave height at breaking point, C_{gb} = wave group celerity at breaking point,
α_b = wave direction at breaking point, $C_{gb} = L_b/T$ (L_b calculated in part a)
$= 44.32/9 = 4.92$ m/s
$\sin 2\alpha_b = \sin(2 \times (-15.59)) = -0.517728399$

The breaker wave height is given by

$$H_b = \frac{H_0}{3.3\left\{\frac{H_0}{L_0}\right\}^{1/3}} = \frac{1.5}{3.3\left\{\frac{1.5}{126.36}\right\}^{1/3}} = 1.9925 \text{ m}$$

$$C_{gb} = n \times C_b, \text{ where, } n = \frac{1}{2}\left(1 + \frac{2kd}{\sinh 2kd}\right),$$

$$K = 2\pi/L_b, \quad K = 2\pi/44.3198 = 0.1417,$$

$$H_b = 0.78 d_b, \quad d_b = H_b/0.78, \quad d_b = 1.9925/0.78 = 2.5545 \text{ m}$$

$$n = \frac{1}{2}\left(1 + \frac{2 \times 0.1418 \times 2.554}{\sinh(2 \times 0.1418 \times 2.554)}\right) = 0.9588$$

$$C_b = L_b/T = 44.32/9 = 4.92 \text{ m/s},$$

$$C_{gb} = n \times C_b = 0.9588 \times 4.92 = 4.7216 \text{ m/s}$$

By Snell's Law,

$$\frac{\sin \alpha_0}{\sin \alpha_b} = \frac{C_0}{C_b}, \quad \alpha_b = \sin^{-1}\left[\frac{14.04}{4.92} \times -0.7660\right] = -15.59°$$

$$P_{ls} = \frac{\rho g}{16} H_{sb}^2 C_{gb} \sin 2\alpha_b = \frac{1024 \times 9.81}{16} \times 1.9925^2 \times 4.92 \times -0.5177$$
$$= -6355.36 \text{ N/s}$$

$$Q = (1290/12) \times (-6355.36) = -683201 \text{ m}^3/\text{month}$$

Similarly for other months the calculations are tabulated below.

Month	H_{sb} (m)	C_b (m/s)	C_{gb} (m/s)	α_b	$\sin 2\alpha_b$	P_{ls} (N/s)	Q (m³/month)
Jan	1.99	4.92442	4.72161	−15.59	−0.5177	−6093.6	−655064
Feb	2.17	5.09486	4.86463	−14.87	−0.4961	−7108	−764110
Mar	1.99	4.92442	4.72161	−11.61	−0.3943	−4640.4	−498847
Apr	1.52	4.30557	4.16858	−4.55	−0.1582	−957.14	−102892
May	1.72	4.76264	4.60501	−5.04	−0.175	−1492.1	−160400
June	3.00	5.98563	5.68265	15.74	0.5222	16822.1	1808373
July	3.24	6.20751	5.86968	17.75	0.5807	22474.2	2415972
Aug	2.59	5.48737	5.23986	10.13	0.34628	7638.81	821171
Sep	2.59	5.48737	5.23986	11.64	0.39522	8718.49	937237
Oct	1.84	4.69854	4.47112	−15.44	−0.5132	−4888.97	−525564
Nov	1.31	4.00011	3.86207	−6.29	−0.2178	−906.91	−97492.7
Dec	1.59	4.41443	4.23269	−8.6	−0.2957	−1980.4	−212889

Direction of drift	m³/yr
Southerly drift	−3017258
Northerly drift	5982754
Gross drift	9000012
Net drift (towards North)	2965496

(b) LEO Method
The detailed calculations for the month of January
The breaker wave height is given by

$$H_b = \frac{H_0}{3.3\left\{\frac{H_0}{L_0}\right\}^{1/3}} = \frac{1.5}{3.3\left\{\frac{1.5}{126.36}\right\}^{1/3}} = 1.9925 \text{ m}$$

$$H_b = 0.78 d_b, \quad d_b = H_b/0.78, \quad d_b = 1.9925/0.78 = 2.5545 \text{ m}$$

Surf width $= d_b \times$ bed slope (1:80 given)

$$= 2.5545 \times 80 = 204.36 \text{ m}$$

$$\left(\frac{V}{V_0}\right)_{LH} = 0.2\left(\frac{X}{W}\right) - 0.714\left(\frac{X}{W}\right)\ln\left(\frac{X}{W}\right)$$

where, $X = 50$ m , W = width of surf zone

$$\left(\frac{V}{V_0}\right)_{LH} = 0.2\left(\frac{50}{204.36}\right) - 0.714\left(\frac{50}{204.36}\right)\ln\left(\frac{50}{204.36}\right) = 0.2949$$

$$P_{ls} = \frac{\rho g H_{sb} W V_{\text{LEO}} C_f}{\frac{5\pi}{2}\left(\frac{V}{V_0}\right)_{LH}}$$

$$= \frac{1024 \times 9.81 \times 1.9925 \times 204.36 \times -0.32 \times 0.01}{\frac{5\pi}{2} \times 0.2949}$$

$$= -5652 \text{ N/s}$$

$$Q = (1290/12) \times -5652 = -607624 \text{ m}^3/\text{month}$$

Similarly for other months the calculations are tabulated below

Month	H_{sb} (m)	W (m)	V_{LEO} (m/s)	$(V/V_0)_{LH}$	P_{ls}(N/s)	Q (m^3/month)
Jan	1.99	204.37	−0.32	0.2949	−5652	−607624.8
Feb	2.17	222.15	−0.31	0.2847	−6702	−720457.1
Mar	1.99	204.37	−0.28	0.2949	−4946	−531671.7
Apr	1.52	155.96	−0.09	0.3245	−841.3	−90434.93
May	1.72	176.12	−0.14	0.312	−1736	−186577.8
June	3.00	308.19	0.4	0.2431	19487	2094893
July	3.24	332.37	0.35	0.2335	20646	2219458
Aug	2.59	265.59	0.19	0.2621	6376.1	685430.4
Sep	2.59	265.59	0.22	0.2621	7382.8	793656.3
Oct	1.84	188.93	−0.38	0.3041	−5562	−597931.4
Nov	1.31	134.4	−0.12	0.3371	−802	−86218.21
Dec	1.59	162.82	−0.18	0.3203	−1858	−199728.2

Direction of drift	m^3/yr
Southernly drift	−3020644
Northernly drift	5793437
Gross drift	8814081
Net drift (towards North)	2772793

(c) **Comparison of energy flux and sediment transport rates computed from above methods**

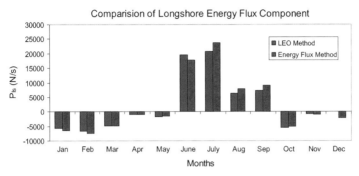

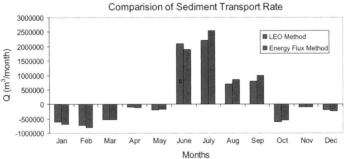

Problem VIII

Compare the sediment transport (m^3) using Kamphuis method, Van Rijn method with CERC method and tabulate.

Month	H_s (m)	T_p (sec)	Deep water direction w.r.t North	Deep water direction w.r.t Shore normal
January	1.5	5	96	29
February	1	5	114	11
March	1	5	156	−31
April	1	5	180	−55
May	1	5	180	−55
June	2	6	174	−49
July	2	6	174	−49
August	2	5	174	−49
September	2	5	174	−49
October	1	5	174	−49
November	1.5	5	153	−28
December	1	5	153	−28

Solution

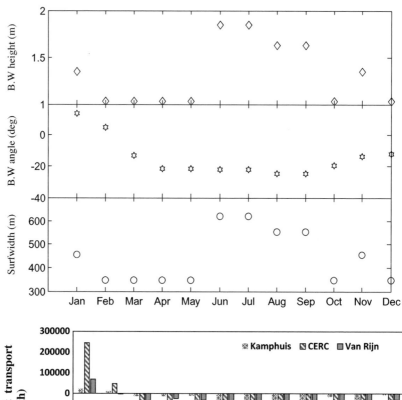

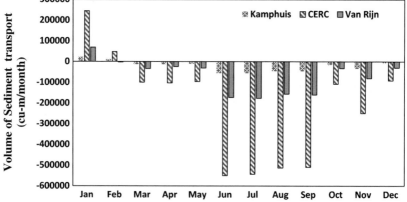

Method	Quantity m³/annum	Ratio with CERC
Kamphuis	−223258	11.32
CERC	−2528218	1.0
VAN Rijn	−781238	3.24

References

Bagnold, R. A. (1966). An approach to the sediment transport problem from general physics, U.S. Geological Survey Professional Paper 422-I.

Battjes, J. A. (1974). Computation of set-up, longshore currents, run-up and overtopping due to wind-generated waves, Rept. No. 74-2, Delft Technical University, Delft, The Netherlands.

Battjes, J. A. (1974). Surf similarity, *Proc. 14th International Conference on Coastal Engineering*, Copenhagen, Denmark, ASCE, pp. 466–480.

Berg, D. W. (1969). Systematic collection of beach data, *Proc. of the 11th Conference Coastal Engineering*, London, September 1968, 273–97.

CERC. (1984). Shore Protection Manual, Vol. I, CERC Dept. of the Army, U.S. Army Corps of Engineers, Washington.

Dorrestein R. (1961). Wave set-up on a beach, *Proc. 2nd. Tech. Conf. on Hurricanes*, Miami Beach, Fla. pp. 230–241. U.S. Dept. of Commerce, Nat. Hurricane Res. Proj. Rep. No. 50.

Kamphuis, J. W. (1991). Alongshore sediment transport rate, *Journal of Waterway, Port, Coastal and Ocean Engineering*, 117, 624–640.

Komar, P. D. (1975). Nearshore currents: generation by obliquely incident waves and longshore variations in breaker height, *Proc. Symp. on Nearshore Sediment Dynamics*. Eds. J.R. Hails and A. Carr, Wiley, London, pp. 17–45.

Komar, P. D. (1979). Beach slope dependence of longshore currents, *Journal of Waterway, Port, Coastal and Ocean Engineering*, Vol. 105, WW4.

Komar, P. D. and Gaughan, M. K. (1972). Airy wave theory and breaker height prediction, *Proc. 13th Coastal Eng. Conf.*, 405–418.

Kraus, N. C. and Larson, M. (1988). Prediction of initial profile adjustment of nourished beaches 10 wave action, *Proc. Beach Preservation Technology '88*, Florida Shore and Beach Preservation Association, 125–137.

Kriebel, D. L. and Dean, R. G. (1985). Numerical simulation of time-dependent beach and dune erosion, *Coastal Engrg.*, 9, 221–245.

Larson, M. and Kraus, N. C. (1989). SBEACH: numerical model for simulating storm induced beach change; Report 1, empirical foundation and model development, Tech. Rep. CERC-89-9. Coastal Engrg. Res. Center, U.S. Army Engr. Waterways Expt. Stn., Vicksburg, MS.

Longuet-Higgins, M. S. (1970). Longshore current generated by obliquely incident sea waves, *J. Geophys. Res.* 75, 6778–6801.

Longuet-Higgins M. S. and Stewart R. W. (1960). Changes in the form of short gravity waves on long waves and tidal currents, *J. Fluid Mech.* 8, 565–583.

Longuet-Higgins M. S. and Stewart R. W. (1962). Radiation stress and mass transport in gravity waves, with application to "surf-beats", *J. Fluid Mech.* 13, 481–504.

Losada, M. A. and Giménez-Curto, L. A. (1982). Mound breakwaters under oblique wave attack; a working hypothesis, *Coastal Engineering*, 6, pp. 83–92.

Mei, C. C. (1983). *The Applied Dynamics of Ocean Surface Waves*, John Wiley & Sons.

Ogawa, Y. and Shuto, N. (1984). Run-up of periodic waves on beaches of non-uniform slope, *Proc. 19th Int. Conf. Coastal Engineering*, pp. 328–344.

Schneider, C. (1981). The littoral environment observation (LEO) data collection program, *Coastal Engg. Tech Aid No. 81-5*, US Army Corps of Engineers, CERC, Fort Belvoir, 24p.

Smith, J. M. and Kraus, N. C. (1990). Laboratory study on macro-features of wave breaking over bars and artificial reefs, Technical Report CERC-90-12, WES, U.S. Army Corps of Engineers.

Sunamura, T. and Horikawa, K. (1974). Two-dimensional shore transformation due to waves, *Proc. 14th Coastal Eng. Conf.*, ASCE, pp. 920–938.

U.S. Army Corps of Engineers (2002). Coastal Engineering Manual (CEM), Engineer Manual 1110-2-1100, U.S. Army Corps of Engineers, Washington, D.C. (6 volumes).

Van Rijn, L. C. (1993). Principles of sediment transport in rivers, estuaries and coastal seas, Aqua Publications, Amsterdam, The Netherlands.

Van Rijn, L. C. (1997). Sediment transport and budget of the central coastal zone of Holland, *Coastal Engineering*, 32, 61–90.

Van Rijn, L. C. (2011). Coastal erosion and control, *Journal of Ocean and Coastal Management*, 30, 1–21.

Walton, T. E. (1980). Littoral sand transport from longshore currents, Technical Note, *Journal of Waterways, Port, Coastal and Ocean Engineering*, 106(WW4), pp. 483–487.

Chapter 4

Coastal Erosion and Protection Measures Including Case Studies

4.1 Introduction

Erosion is said to occur along a coast when the quantity of loss of sediments surpasses the quantity supplied to it. Any beach is supposed to be in equilibrium, when the sediment movement for a period of a few years is considered. However, some beaches experience excessive erosion or advancement of shoreline disturbing the equilibrium which is the resultant of wave climate and morphology and is location specific.

Coastal erosion is a common problem along the coasts around the globe calling for its protection that include cultivate lands, valuable properties, power plants, tourist attractions, etc. The erosion resulting in the loss of beaches and significant portion of the shoreline is common during storms, cyclones and tsunami. The loss of beach material along the shore is usually sudden and devastating. Apart from the said causes due to the said extreme events that would last only for a few days, the perennial erosion along a coast would be due to natural and manmade or a combination of both. In this case, coastal erosion can also be gradual process over years even without any intervention through structures. The causes for erosion need to be thoroughly examined prior to the proposal of a solution or a protection measure. The manmade causes would be due to the construction of structures like breakwaters for development harbours, or shore connected solid or rubble mound structures jetting into the sea that are installed for either drawing sea water into land for aquaculture, power plants, etc., the effects of which are known priori if careful modeling of sediment and wave dynamics is carried out in the planning stage.

4.2 Erosion Process

The wave that generates in the offshore propagates towards the coast experiencing the phenomena of shoaling, refraction, diffraction and combined effects. They steepen due to the shallowing of the bottom depth contours and finally break. When the incoming wave from deep water breaks, its energy is dissipated along the surf zone width and when this width is insufficient or less to dissipate the energy the remaining energy is spent in eroding the sediments. These eroded materials either get suspended in the littoral currents or gets deposited in the nearshore bottom forming an offshore bar. When the offshore bar becomes large enough to facilitate breaking of incoming waves, the energy contained in them gets dissipated in the surf zone. The above said process of shore erosion due to the attack of storms is illustrated in Fig. 4.1. In addition, at locations dominated by alongshore currents moving the sediments, man-made or natural barriers in the near shore obstruct the free passage that would lead to advance of the shoreline on the updrift side and erosion along the shoreline on the downdrift side.

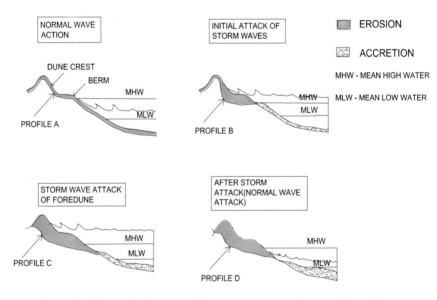

Fig. 4.1 Schematic diagram of storm wave attack on beach and dune.

4.3 Causes for Coastal Erosion

The broad classification of the causes for erosion due to nature and man-made obstructions are brought out in Table 4.1.

Table 4.1. Causes for coastal erosion.

Nature	Man-made
Rise in sea level	Construction of dams, dykes and other coastal structures.
Protruding headlands, reefs or rocks into the sea	Groins, breakwaters, jetties etc.
Tidal entrances and river mouths causing interruption of free passage of sediments along the shore, natural protection of tidal entrances	Manmade entrances causing interruption of littoral drift. This includes construction jetties.
Shoreline geometry causing rapid increase of drift quantity	Fills protruding in the ocean to an extent that they change local shoreline geometry radically.
Removal of beach material by wind drift	Removal of material from beaches for construction and other purposes.
Removal of beach material by sudden outbursts of flood waters	Digging or dredging of new inlets, channels and entrances offshore dumping of materials.

4.4 Strategy for Coastal Protection

Prior to planning of protection measures, one has to understand the basic behaviour of waves in moving the sediments and its interaction with sediments that decides the zone of erosion and accression. The block diagram shown in Fig. 4.2, clearly explains the conditions under which a coast could be stable or otherwise.

Whenever erosion occurs, there are some guidelines to be followed. Coping erosion along a sandy coast is different from coping erosion along muddy coast, mangrove coast, and coast with clay or rock. The following are the suggested procedures:

- Verify if the erosion is temporary due to seasonal effect.
- Work out the cost of different alternatives. The costs should include not only maintenance, construction etc. but also in terms of loss of cultural values, impact on safety and the needs of the local public, etc.
- Fund/resources to combat erosion in a sustainable way.

Only after ascertaining the above basic resources and needs, proceed as follows:

- If fund/resources are available, then combat erosion permanently by proper planning.

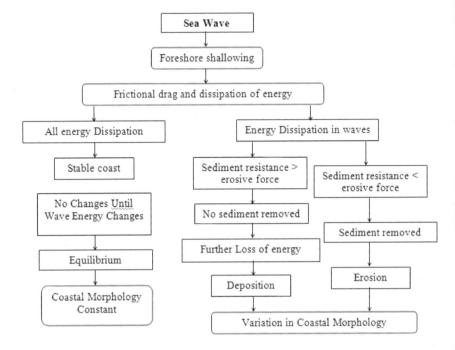

Fig. 4.2 Interaction of waves and sediments.

- If enough funds or resources are unavailable, careful planning of temporary measures is essential.

Requirements for a detailed evaluation for protection measures:

- Collection of seasonal field data and analyze the same critically.
- Use old and new satellite imageries to assess the shoreline behaviour.
- Use of G.I.S as a tool to map the coastal region. This would help in the planning process of coastal protection.
- A field visit along the coast.
- If erosion is observed continuously over a number of years, it is chronic erosion.
- If a coast is stable over a long period, but subjected to occasional severe erosion (due to cyclone etc.) and then recovers, it is called acute erosion.
- The effect of the recent tsunami on the shoreline should also be taken into account while detail planning is taken up. This can be accomplished using the techniques of remote sensing and G.I.S.

Furthermore, prior to looking at the options in hand, for a coast under distress due to erosion, a coastal sediment budget will help in the planning of the protection measure. The options for coastal protection are listed below.

- Do nothing.
- Remove the causes for coastal erosion although is the best solution but always may not feasible.
- Supply sediments to the affected area termed as artificial nourishment.
- Reduce loads by constructing submerged breakwaters in front of the coast.
- Increase the strength by constructing shore defense structures (hard measure to be discussed later).
- Vegetation termed as bio-shields.

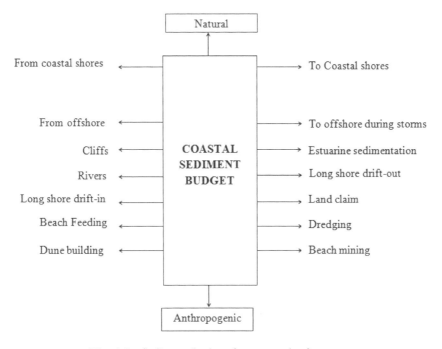

Fig. 4.3   Sediment budget for a stretch of a coast.

The basic physical factors needed for selection for the selection of coastal protection measure are geomorphology of the coast, material characteristics and sources, characteristics of wave, currents, tides, winds and storms, shoreline details, bathymetry, shoreline oscillations over the past years, and most importantly the magnitude and direction of littoral drift.

4.5 Coastal Protection Measures

4.5.1 *General*

Coastal protection is any manmade measure or structure in the vicinity of the shoreline, (i.e. between the shoreline and breaker line) to combat against the forces that bring up a drastic change in the coastal morphology. This could be either protecting the coast as well as winning the beach through deposition of sand through artificial means or by construction of structures. The other situation calling for protection is for creating desired erosion or avoiding/minimizing siltation in particular near river mouths. Protection measures can be broadly classified as hard and soft measures. The detailed classification is shown in Fig. 4.4.

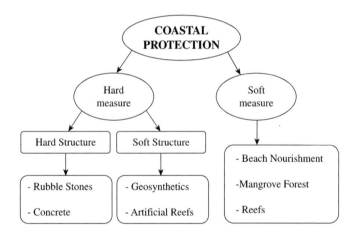

Fig. 4.4 Broad classification of coastal protection measures.

4.5.2 *Hard measures*

This refers to the protection measures that will leave behind a morphological/geographical/geological/environmental/ecological foot print in addition to the shore protection itself. In other words, a permanent change is effected as a result of any structure or measure undertaken to improve the stability of the shoreline. These measures when designed and executed properly are most likely to yield desirable results in a shorter duration of time. However, improper designed and unskilled execution might end up worsening the issue in hand.

4.5.2.1 *Hard structures*

Hard structures are solid and sturdy structures usually made up materials like rocks/rubble and concrete. Most of the coastal protection schemes involve construction of such hard structures. The various hard structures in practice are discussed in detail below.

Offshore breakwaters

The main purposes of breakwaters are for the formation of artificial harbors, to protect an area inside the sea against waves, to reduce dredging at harbor entrance, to serve as quay facility, to guide currents, to provide tranquility conditions inside the harbor. Offshore detached breakwaters are a widely adopted option for protecting the coast against erosion along a littoral drift dominated coast and they have proven to be effective in reducing and absorbing the incident wave energy. The waves around such structures undergo transformations like, shoaling, refraction, diffraction reflection from structures as well as the combined effects of all these phenomena. Their cross-section is usually of mound type and are aligned parallel to the shore. These structures protect the shore by preventing the direct attack of destructive waves. The advancement of the shoreline until it reaches the breakwater is termed as "Salient" and when it reaches the breakwater; it is termed as "tombolo".

As the construction of offshore breakwaters requires skilled labor and heavier machineries and quite expensive compared to shore connected structures, it should be considered only after ascertaining its effect as an effective protection through detailed investigations before implementation. The shoreline variation due to construction of a single offshore breakwater and due to segmented offshore breakwaters is shown in Figs. 4.5(a) and 4.5(b), respectively. Usually, the offshore breakwaters are of exposed type, and of late, submerged type termed as multipurpose artificial reefs have become popular.

Hanson and Kraus [1990] indicated that in general there are at least fourteen parameters, including breakwater parameters, hydraulic conditions, and sediment properties which control the shoreline response of a sandy beach: The breakwater parameters are length of structure (LB);

distance from structure to the original shoreline (XB); gap width between structures (GB); structure transmissivity (K_t); and orientation of structure to the original shore (θ_B). The hydraulic conditions include wave characteristics (height (H_s); period (T); predominant wave angle (θ) whereas, the water level include water depth at structure and tidal range. Apart from the above, the sediment properties are represented by the sand median size (D_{50}). There have been considerable numerical and physical model studies from which the general behaviour of the shoreline depending on the variations in the above stated parameters could be ascertained. Pictorial representations of the parameters that govern the formation of salient/tombolo are presented in Fig. 4.5(c).

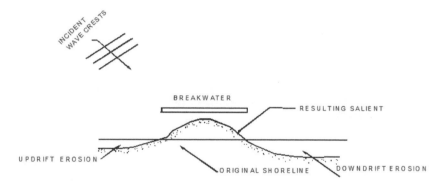

Fig. 4.5(a)   Shoreline evolution due to a single offshore breakwater.

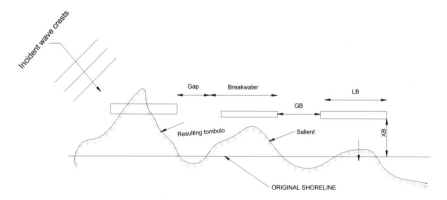

Fig. 4.5(b)   Shoreline evolution due to segmented offshore breakwater.

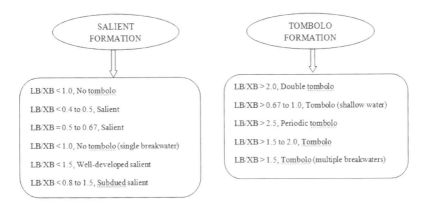

Fig. 4.5(c) Formations of salient and tombolo based on the environmental parameters.

Submerged offshore breakwaters function as dissipaters reducing the incident wave energy and the waves undergo diffraction. If planned for, it can also potentially bring down overtopping and run-up on coastal structures. The submerged structure facilitates premature breaking for certain wave characteristics. A comparison of coastal process in the presence and absence of submerged breakwaters is illustrated in Fig. 4.5(d).

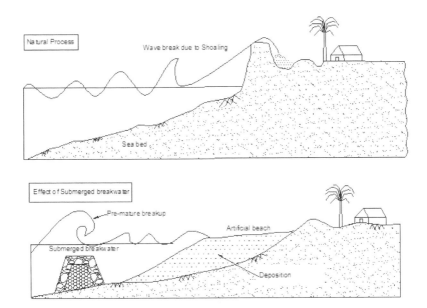

Fig. 4.5(d) Coastal processes in the absence and presence of submerged breakwaters.

Seawalls and bulkheads

A seawall is a structure constructed parallel to the coastline that shelters the shore from wave action. It is both massive and expensive. Erection of sea wall shall be considered only after carefully evaluating the socio-economic parameters, i.e. it is recommended to construct a seawall when the adjoining shore is highly developed and the intensity of storm attack is severe. It is an irreversible process where the beach area in front of the structure is permanently lost. Seawalls are broadly classified as rigid, flexible and composite type. Rigid seawalls are the structures generally most of its seaward side face being vertical, sloping or curved and are impermeable. These seawalls are usually constructed as monolithic gravity structures consisting of a concrete caisson filled with sand. It provides complete protection for the shore from the destructive action of ocean waves, and is provided with a toe protection to prevent toe scouring. The flexible seawalls are permeable rubble mound structures armoured with natural rocks or artificial armour units or gabions. The dissipation of the incident wave energy in the presence of these structures is mainly due to wave breaking on the slope. Rehabilitation of the lost armour layer units is mandatory for longevity of the structures. Composite seawalls are a combination of rubble mound seawall with a crown of rigid impermeable wall (either vertical or curved), wherein, the crown wall rests on the mound of stones, relying on its mass to resist the wave action. This type is adopted at locations, where, the foundation soil is poor and a high tidal variation pertains. During low tide, the structure will act as rubble mound (permeable type) and during high tide it acts as impermeable wall type, with less force on the rigid impermeable wall due to the presence of rubble mound toe in front. Some special types of seawalls are also being constructed using gabion mattress and geotubes filled with sand. The different types of seawalls as discussed are illustrated in Fig. 4.6. A bulkhead is very much similar to a seawall except for its primary purpose which is to prevent sliding of land mass, whereas, a seawall's primary purpose is to protect backshore against wave action.

Groins

A groin is a shore connected structure constructed perpendicular to the coastline and jetting into the sea which serve as barrier in trapping the longshore sediment transport or control longshore currents. Different plan

4. Coastal Erosion and Protection Measures Including Case Studies 107

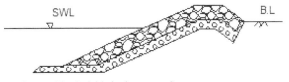

Permeable straight sloping seawall

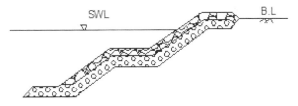

Permeable sloped seawall with berm

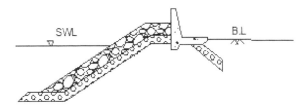

Permeable sloping seawall with an impermeable crown wall

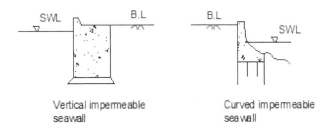

Vertical impermeable seawall

Curved impermeable seawall

Fig. 4.6 Types of seawall sections [CEM, 2002].

shapes of groins have also been employed for trapping the sediments in the nearshore (Fig. 4.7(a)). The major parameters that decide the effectiveness of a groin in a given location are net and gross sediment transport rates, water depths at the seaward tip of the groin, surf zone width, structural permeability and grain size distribution of sediments. A number of closely spaced groins, termed as "groin field" will tame the movement of sediments along the coast which is likely to be trapped in between the gaps (Fig. 4.7(b)) which over the years will result in winning the lost beach which is its added advantage over a seawall. The oldest groin field was constructed in Vissingen, The Netherlands in 1503. It can be observed that as an effect of groin construction deposition or shoreline advancement takes places on the updrift side and erosion or shoreline recession takes place on the down drift side. The length of the individual groin should be beyond the surf zone width of the given beach, the spacing between two groins in a groin field should be 1.5 to 3 times the length of the groin. Although, the alignment of the groins is usually perpendicular to shore, it can also be inclined at a specific angle for functional purposes. Timber, steel, stone, concrete, geosynthetic materials or a combination of them are adopted for construction which is site dependant. The groins usually penetrate the free surface; however at locations dominated by bed load sediments, submerged may be adopted. The lengths and spacing need to be carefully finalized through numerical or physical modeling, it is extremely important that the groin spacing helps in the preservation of beach levels and act as active trappers of the sediments in motion along the shore. If the spacing is more, the retention of the sediments is poor as illustrated in Fig. 4.7(c), whereas, transitional groin field as a means of reducing impact of groins on downdrift coastlines (Fig. 4.7(d)). The phenomena of shifting of the erosion to the adjoining shoreline is minimized or avoided. If the spacing between the groins is too small loss of trapped sediments towards offshore due to rip currents is likely to take place.

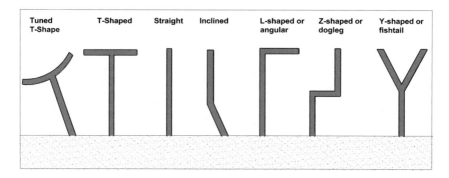

Fig. 4.7(a)   Different plan shapes of groins.

4. Coastal Erosion and Protection Measures Including Case Studies   109

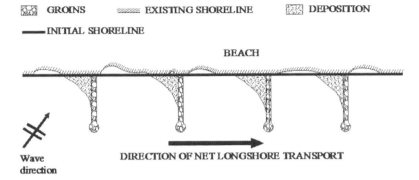

Fig. 4.7(b)   Typical shoreline evolution due to a groin field.

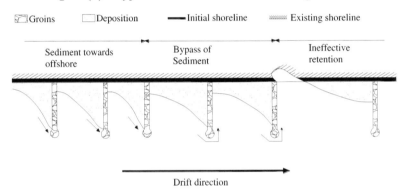

Fig. 4.7(c)   Effect of spacing between groin field (poor retention for increased spacing).

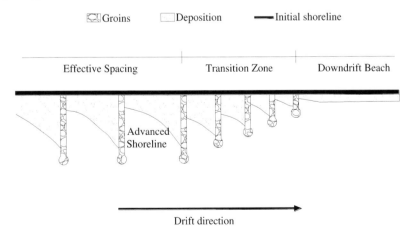

Fig. 4.7(d)   Effect of transition groin field (progressive reduction in the deposition of sediments).

Jetties

Jetties are usually constructed perpendicular to the shore; it can be effectively used inside harbor basins for transfer of cargo. It can be broadly classified as open jetty (piled) or closed/solid jetty (closed walls with earth fill). The solid jetties can influence the shoreline dynamics similar to that of a groin. Figure 4.8 shows a typical layout of jetty. It demonstrates that this is longshore drift dominated coast that has lead to the advancement of the shoreline on the updrift, whereas, the downdrift side is protected by a groin field as a countermeasure for erosion. Groin is also spelt as groynes.

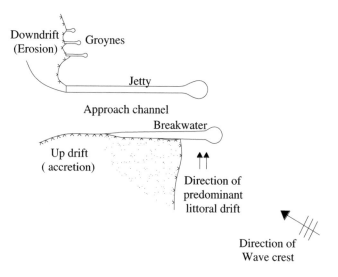

Fig. 4.8   Layout of a jetty.

Revetments

Revetments are structures deployed primarily to improve the stability of the sloping surfaces along the shore. It is a sought after durable and environment friendly erosion protection system for berms, slopes, riverbanks and shorelines. The revetments can be made up rubble stones, concrete blocks with good interlocking capacity or even geosynthetics and geosystems. Figure 4.9 shows a typical cross-section of quarry stone revetment, which is quite easy to construct. Interlocking concrete blocks of different shapes also exist for forming the revetment.

4. Coastal Erosion and Protection Measures Including Case Studies   111

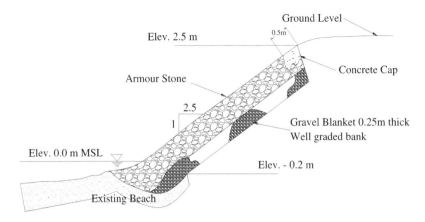

Fig. 4.9   Cross-section of quarry stone revetment.

Training walls

They are usually constructed at tidal inlets and as the name suggests their primary purpose is to train the river mouths. These structures propagate from the river mouth into the sea, facilitating tidal flushing; preventing sand bar formation and clogging of river mouths. Typical layouts showing the effect of training walls at the mouth of Perumanathura (8°37′59.41″N and 76°47′11.98″E) along the southwest coast of India are shown in Fig. 4.10.

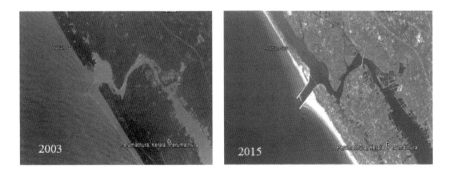

Fig. 4.10   Google earth imagery of Perumanathura.

4.5.2.2   *Soft structures*

Soft engineering options dissipate incident wave energy, the extent of which depends on the nature of the protection measure, wave and sediment dynamics along the coast. These include structures reinforced with eco-friendly materials which do not greatly impact the environment and eco-system.

Geosynthetics

The Bureau of Indian Standards have defined geosynthetics "all synthetic materials used in geotechnical engineering applications, which includes geotextiles, geogrids, geomembranes, and geocomposites" [IS 13321-1992 (Part 1)].

A more elaborate definition to geosynthetics is given by the American Society for Testing and Materials (ASTM) Committee D35 on geosynthetics as *"a planar product manufactured from polymeric material used with soil, rock, earth, or other geotechnical engineering related material as an integral part of a man-made project, structure, or system."*

In general, geosynthetics are manmade materials used to improve soil conditions. The word is derived from: *Geo* = earth or soil + *Synthetics* = manmade. Geosynthetics have proven to be among the most versatile and cost-effective ground modification materials. They are synthetic products used to solve geotechnical problems. The development of geosynthetic applications in geotechnical and hydraulic engineering has been very rapid. In the beginning the terms filter mats and woven geotextiles were used for all water-permeable geosynthetics, and the term membrane was used for all water-impermeable geosynthetics. The classification of geosynthetics is presented in Fig. 4.11.

As water-permeable geosynthetics, geotextiles are suitable for filtering, drainage and separation of soils with different grain compositions. Because the fibers and the geotextiles products possess certain strength, geotextiles can resist and transfer forces. They are therefore suitable for performing the tasks of reinforcement and encapsulation and can also serve as a protection layer. "Building appropriate to geosynthetics" often means, that it is impossible for a special application, to initially define a single function of the geotextiles and design according to it.

The following factors should be taken into consideration when dimensioning and selecting geotextiles.

- Grain-size distribution/grain-size spectrum of the soil to be filtered.
- Where necessary, the plasticity index for cohesive soils.
- The hydraulic loads (hydrostatic/hydrodynamic magnitude of the hydraulic gradient).

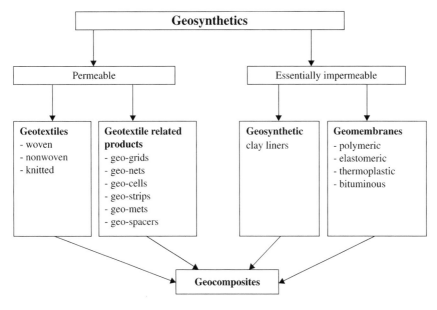

Fig. 4.11 Classification geosynthetic products.

- Loading case (orientation of the contact surfaces/flow direction).
- Type and method of construction of the structure, where necessary.
- Structure-related safety requirements posed on the filter and
- Installation loads.

A comprehensive discussion on the characteristics of geosynthetics and its application in the marine environment are discussed in detail by Pilarczyk [2000].

Artificial reefs

Reefs are naturally formed bars of rock, sand or corals underneath the ocean water surface. Coral reefs are often known as the rainforests of the sea. Reef formation can be a result of either biotic or abiotic processes.

Artificial reefs are manmade underwater obstructions that are submerged below the sea surface. These obstructions are like any natural ones acting as buffers in diffracting the incident wave energy, offer friction, thus leading to wave attenuation. The degree of attenuation of incident wave energy depends on the shape, size and material of the artificial reefs. The artificial reefs are also expected to enhance recreational benefits like diving, fishing and surfing. The geometry and design of the artificial reefs varies as it depends on the main functions of the structure and materials used. All kinds of materials such as steel, reinforced or pre-stressed concrete, fiber glass or a variety of composite materials have been used in construction of artificial reefs. Old wrecked cars, airplanes, military tanks, used truck or car tires, junked appliances, docks, old boats, ballistic missiles, decommissioned ships and obsolete oil rigs have been sunk and designated as artificial reefs.

4.5.3 Soft measures

It refers to the protection measures which leaves behind less or no morphological/ geographical/geological/environmental/ecological foot print. These measures aim to stabilize the shoreline and save the beaches and landforms from any approaching natural disasters. These measures are time consuming to both plan and implement. It requires careful planning and skilled labor for execution. Although the efficiency of such measures are commendable, it is not a quick solution.

4.5.3.1 Beach nourishment

General

A wider beach width can be artificially achieved by supplying sediments from a source to the stretch of the coast facing erosion. This methods aids to enhance the recreational value of the beach further acts as a buffer to dissipate high energy waves. Wider beaches are often sought after not only for recreational or tourism purposes; they can potentially withstand damage caused by heavy storms, tsunami, storm surge, etc. and aid in energy dissipating along the surf zone. The source for sediment for artificial nourishment shall be collected either from offshore dredging operations or alternative landward sources, of which, the marine source of sediments is preferable. Artificial nourishment requires continued maintenance and in likely to become uneconomical in severe wave climates. This is often employed in combination with hard structural/engineering options, i.e. offshore breakwaters, headlands and groins to improve efficiency.

The primary advantage of this method is that it is more economical than deploying huge defense structures. Nourishment can broadly be classified as backshore nourishment, beach nourishment and shore face nourishment (Fig. 4.12).

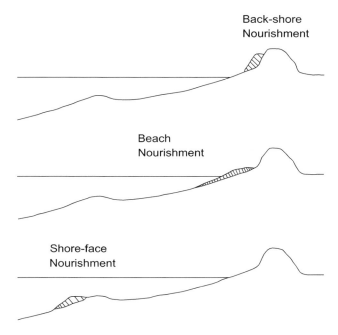

Fig. 4.12 Principles of beach nourishment.

(a) *Backshore*

Backshore nourishment usually feeds sediments/sands into the uppermost region of the profile i.e. on the foot of dunes or backshores. This aims to mitigate the erosion/breaching of dunes against extreme events, and ends up sacrificing huge quantities restored sediments in the front face during the same. This method can be viewed as an emergency measure to prevent dune setback during storms.

(b) *Beach*

The supply of sand to the region of eroded beach intended to increase beach width and thereby its recreational value is known as beach nourishment. This is not a direct coastal protection measure, although a wider beach acts as a buffer to withstand any possible extreme event or even disaster. It is to be noted that the borrowed material used for replenishment must be analogous with the serving/parent sediment, such as to adjust to the existing profile. Use of slightly coarser sediment than the serving/parent sediment is favourable, as it improves the stability by resulting in the formation of a slightly steeper profile, whereas, fine sands are easily suspended and relocated.

(c) *Shore face*

Shore face nourishment is the supply of sand to the lower part of the beach, especially on the sea side of an existing bar. This measure is most suited for regions with long term sediment deficiency or in areas where coastal protection schemes are implemented. Shore face and beach nourishment can be coupled to yield strengthened coastal profile.

4.5.3.2 *Placement of sand and borrow site*

- The sand could be dumped in the affected area to take the natural profile. The dredged spoil is discharged inshore through a split hopper dredger or directly pumping directly over to the beach through pipelines. Though, the distribution of size of the sediment and its influence in the beach nourishment was suggested by few researchers, including Shore Protection Method. Sands on the updrift side, that are intercepted by nature (river mouths, approach channels, etc.) or man-made structures, from nearby estuaries, lagoons, backwaters, inlets can serve as sources for beach nourishment. The main drawbacks are:
- Once placed on the beach, the quality of the beach depends on the quality of sand adopted for nourishment which is often not fulfilled.
- Source for quality sand become quite difficult at certain locations. Continuous exploitation of the source would lead to depletion.
- Dredging of sand is not eco-friendly to eco-system in particular to marine life like the sea turtles.
- It is cost intensive and involves regular maintenance. For instance, the pumps are designed with a certain handling capacity. During the monsoon, when the sediments to handle are significant, power failure would be a drawback. At this instance, dredging would be suspended leading to a steep cumulative increase in the quantity to be dredged. This needs careful attention.

4.5.3.3 *Methods*

There are two major design methods for the design of the beach nourishment, one is mathematical erosion model, which works only in case erosion is clearly caused by a gradient in long shore transport. The uncertainties in this method are quite high.

4. Coastal Erosion and Protection Measures Including Case Studies

The other method, i.e., the Dutch Design, discussed by Verhagen [1992] is widely adopted in many European countries. Based on this method, the following basic steps should be considered for designing a beach nourishment project.

- Perform coastal measurements (preferably for a decade).
- Calculate the "loss of sand" in m^3/year per coastal section.
- Add 40% loss.
- Multiply this quantity with a convenient lifetime (say, five years).
- Place this quantity somewhere on the beach between the low-water-minus-1-meter line and the dune foot.

This method is simple and straightforward and does not require mathematical models and wave (or wind) data, but good quality measurements of the position of the coastline are essential. The above method is based on the assumption that the beach nourishment has no influence on the long-term natural behaviour of the coast. Or in other words, the erosion rate before nourishment equals the erosion rate after nourishment [Verhagen, 1992].

There are four main methods for artificial beach nourishment namely, offshore Dredging, Direct Placement, Bypass System and trickle Charge.

The offshore dredging method is dredging the sand in the offshore by a cutter-suction or a trailing suction hopper dredger and placed on the affected beach that has been experiencing erosion due to storms or wave action. This method can be accomplished either by Rainbow Pump Method or pipelines Linked Method. The rainbow method being commonly adopted in which, a hopper dredge is usually used to dredge and retain it as slurry as the sand required is to be sprayed onto the required affected area to be replenished through a nozzle. Since the path of the sprayed slurry resembles a rainbow it has derived the name. The pipelines linked method is adopted at locations where the dredgers do not have enough draft to dispose the dredged spoil from the ocean. The pipeline is extended from a dredger which is moored at a certain distance from the shoreline. This type of dredging system can be used with cutter suction dredgers as well as hopper suction dredgers. The pipeline can be mounted on a trestle or on a floating barge.

The direct placement method is the transfer of sand the characteristics of which being quite similar to that at the site to be replenished. This is achieved by mechanical means through conveyor belts, bulldozers, etc., from identified sources either from offshore or from land. At locations structures intercepting the free passage of the littoral drift alongshore, the sand from the updrift can serve as a source to nourish the beach on the downdrift side.

The sand bypass system involves dredging from locations of deposition to the eroding areas. This usually is the best measure to maintain the opening of an inlet by preventing the sediments closing its mouth or in the event of harbour breakwaters acting as littoral barriers facilitating the advancement of shoreline on the updrift side and erosion on the downdrift side.

In the trickle charge method, the beach is recharged gradually by depositing the sediments close to the low tide level or beyond close to the beach, which could be driven by the waves and currents towards the shore the beach. This usually results in the formation of a natural beach.

4.5.3.4 Vegetation cover

After the 2004 tsunami there was extensive study of the aftermath of the tsunami and about various means to mitigate the effects of such catastrophes. A common observation in such studies was that many areas with thick vegetation at coast were not much affected by the tsunami. This has lead to a suggestion of using trees and coastal vegetation as a means to protect the coast. However, there are views in support and against this measure. The following are some of the testimonials in support and against bio-shield as coastal protection measure. The vegetation beyond any doubt helped to protect against tsunami. The spacing between the vegetation is an important factor in deciding the level of protection offered by the vegetation. The trunk diameter and the spacing between the trees have a close relationship with each other. The vertical distribution of the drag coefficient varied identifiably with species classification.

Field observations and laboratory research have established several key parameters that determine the magnitude of tsunami mitigation offered by various types of coastal forests. These parameters include:

- Forest width,
- Tree density,
- Age, tree diameter,
- Tree height, and
- Species composition.

Based on studies and scientific results, the presence of vegetation in coastal areas improves slope stability, consolidates sediment and reduces wave energy moving onshore; therefore, it protects the shoreline from erosion. However, its site-specificity means that it may be successful in estuarine conditions (low energy environment), but not on the open coast (high energy environment). In some cases, vegetation cover fails because environmental conditions do not favour the growth of species at the particular site or there is ignorance as to how to plant properly given the same conditions. It is also possible that anthropogenic influences have completely altered the natural processes in the area. The most obvious indicator of site suitability is the presence of vegetation already growing. This can be extended by other factors such as the slope, elevation, tidal range, salinity, substrate and hydrology [Clark, 1995 and French, 2001].

Mangrove forests are dense plantation that grows in seawater. These plantations have specially adapted root system, which goes deep inside the water bed. The roots get tangled and almost the trunk the submerged in water. The roots help in holding the ground intact and prevent soil erosion due to wave action. Also, due to dense nature, the wave attenuation is very high. The ability of waves to carry sediments is proportional to its flow velocity, which is drastically reduces on incidence with mangrove forests and facilitates sediment settlement, soil build up and consolidation of land.

4.6 Case Studies

4.6.1 *Concept generation*

Although, several stretches of the coast around the globe are subjected to severe erosion leading to instability in the shoreline, the two maritime states of India, namely Tamil Nadu and Kerala that have been experiencing perennial erosion as well as suffered severe damage during the great Indian Ocean tsunami of 2004 are considered herein for case studies. In order to protect a few stretches vulnerable coast from severe erosion, hard measures such as seawall and groin field are proposed and, many of them have been executed. The planning of the protection scheme commenced from reconnaissance survey from identification of the vulnerable stretches as well as those affected during extreme events like a cyclone to field monitoring of beach profile, bathymetry up to twice surf zone width and sediment characteristics. Further, the environmental forcing such as wave and current which drive the sediments had to be collected and evaluated near shore. With the predicted the mean breaker height, breaker wave angle, surf width and net drift magnitude the design of protection layout were finalised. The proposed shore protection structures are subjected to numerical shoreline evolution to predict the shoreline patterns after constructing the groin field. The shoreline patterns predict the shore advancement and erosion say after 1, 5, 10, 15, 20 and 25 years through numerical modelling.

The coastal activities in Tamil Nadu have been accelerating due to industrial growth and tourism in addition to the primary habitants in the region. On the other hand, the population density near the coast of Kerala state is many times higher compared to its inner region. The majority of the community residing along the coastal belt mainly depends on fishing related works. The coast has been experiencing perennial erosion and during 1950s to 1970s, nearly 400 km stretch of coastline had been protected with seawalls of a standard cross-section, which was not location specific. After the tsunami of 2004, the coast has been subjected to severe attack of waves resulting in the instability of the coast, choking of river mouths and land loss due to erosion. Master plans for coastal protection for maritime states of Tamil Nadu and Kerala affected by tsunami 2004 was prepared by mid-2005 by the authors. Although, several of them have been implemented only a few examples are presented and discussed in this chapter. Figure 4.13 shows the geographical location of the states of Tamil Nadu and Kerala.

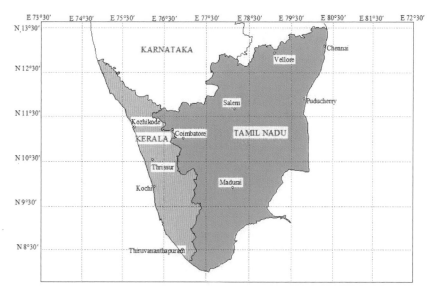

Fig. 4.13 Geographical location of the states of Tamil Nadu and Kerala in the southern Indian peninsula.

4.6.2 Tamil Nadu (groin field)

4.6.2.1 North of Chennai Harbour (transitional groin field)

The port of Chennai is located along the east coast of Indian peninsula (13.0844°N, 80.2899°E). Ever since, the formation of harbour of Chennai port with breakwaters, the coast on its north has been subjected to erosion at a rate of about 8 m per year due to the predominant northerly drift. A part of the existing National Highway and the residential area nearer to this coastline has already been sacrificed to the sea. In spite of the provision of a seawall, the erosion continued along few pockets along the coast and the severely affected zone is shown in Fig. 4.14. The location of the study area as well as another port in Ennore north of the Chennai port commissioned in 2001 is also indicated in this figure.

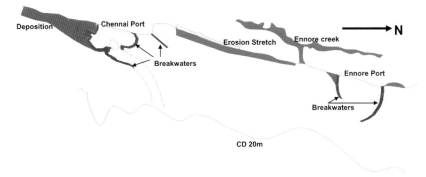

Fig. 4.14 Schematic view of the affected area in between Chennai and Ennore ports.

The measures adopted over the past four decades did not solve the problem of coastal erosion. A highway being not only the main link between the ports of Chennai and Ennore but also serving as the only link to several industries along the coastal city has been experiencing severe traffic congestion, many times to a stand still for hours together mainly due to the sacrifice of part of the highway to the Bay of Bengal explains the importance in protecting this stretch of the coast. The solution of a groin field to combat the erosion problem and the success of the project are highlighted herein.

The solution for the coastal erosion problem was divided into two categories, a temporary strengthening of the existing seawall and a permanent remedial measure by providing suitable groins. In the first phase, a detailed bathymetry survey for the measurement of existing cross-section of seawall and its status in order to assess its adequacy for the design wave climate were carried out. The wave characteristics such as average wave height, wave period and wave direction, from which the average breaking wave characteristics were derived from different sources. The monthly sediment transport has been estimated based on Energy Flux method [CEM, 2002], the method of Komar [1969] and by integrating the distribution of sediments within the surf zone. The net sediment drift along the Chennai coast is observed to be about 0.8×10^6 m^3/year directed towards the North. In the second phase, as a permanent solution for the coastal erosion problem, ten numbers of shore-connected straight rubble mound groins in the two severely affected stretches (stretch I and II) shown in Fig. 4.15 were proposed. The length and the spacing between groins were designed based on the recommendations of coastal engineering manual [2006], the details of which are projected in Figs. 4.16(a) and 4.16(b). Mathematical modeling to evaluate the shoreline changes due to the proposed groin field was carried out. The mathematical modeling of shoreline evolution essentially relates the change in the beach volume to the rate of material transported from the beach. The methodology for the present numerical model is based on the numerical scheme proposed by Janardanan and Sundar [1997] and the one line model solved by using Crank Nicholson implicit finite difference method. The construction of the proposed groin field started in May 2004. Immediate shoreline advancement on the south of the executed groin has been substantiating the most favourable choice and design of the suggested remedial measure. The approximate beach widths formed due to the groins 6 and 5 are shown in Fig. 4.17.

The tsunami run-up has initially taken away nearly 50% of the beach that was formed in between the groins and within a short duration of time slight more than what was removed has been re-deposited. The area of the beach obtained through continuous monitoring for the different periods are shown in Table 4.2. The details of this study are discussed by Sundar [2005].

4. Coastal Erosion and Protection Measures Including Case Studies 123

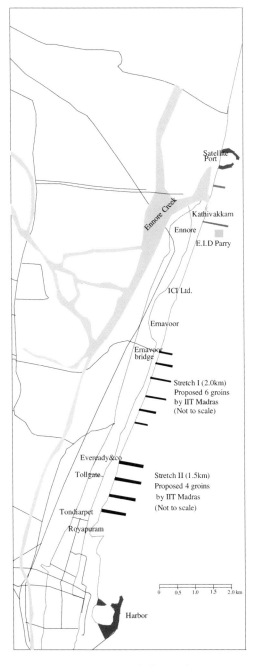

Fig. 4.15 Layout of the study area.

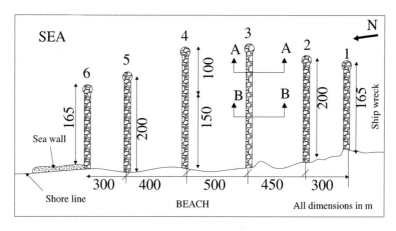

Fig. 4.16(a)   Layout of groin field for stretch I.

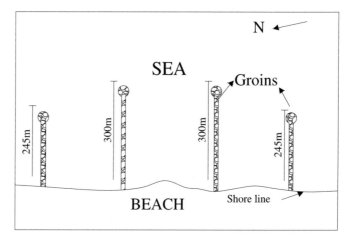

Fig. 4.16(b)   Layout of groin field for stretch II.

4.6.2.2 *Vaan Island (submerged artificial reefs)*

There is a rich content of coral structure available near the Vaan and Koswari Islands (8.83639°N, 78.21047°E). Coral reefs are often known as the rainforests of the sea. It is known to have high productivity in coastlines. The protective measures are planned in such a way that the structure should minimize wave impacts particularly storms such as cyclones as well as serve as a suitable substrate for natural coral recruitment. Hence, once the Coral recruits are attached in this platform in the future, due to its

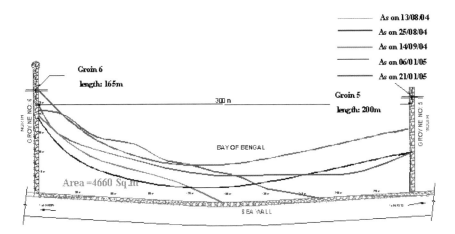

Fig. 4.17   Shoreline advance in between groins 5 and 6 for different periods.

Table 4.2.   Area of beach in between groins 5 and 6.

Date of Measurement	Area in m^2
Work commenced in May 2004	
13 Aug'04	3700
25 Aug'04	6970
14 Sep'04	8800
Post Tsunami	
06 Jan'05	4660
21 Jan'05	10450

rough surfaces and complex structures, these dissipate much of the force of incoming waves. The locations of these two islands north of the Tuticorin port are shown in Fig. 4.18(a). The Vaan Island which is located off Tuticorin coastline of Gulf of Mannar has shown substantial erosion after 2010. This might be due to many environmental factors and other threats of degradation of coral reefs surrounding it. The length of the island along north-south was about 1 km till early 2013. However, the northern part of the island has been disintegrated during late 2013 and at present, the length is about only 300 m. It is observed that there are submerged sand deposits on the northern part of the island. These deposits span along east-west direction, indicating that the island could have been disintegrated during north-east monsoon wave climate. The main objective of the shore protection for this island is to design suitable protective measures for the

126 Coastal Engineering: Theory and Practice

Vaan Island by keeping in mind that it should not have any influence on the neighborhood island, Koswari, and, it should facilitate the growth of corals. An analysis of the wave climate in this region is required in order to prescribe suitable protection measures. For this purpose, a numerical model developed to consider the combined refraction-diffraction small amplitude surface gravity waves over an arbitrarily varying mild sloped sea bed as proposed by Berkhoff [1972] also discussed by Li [1994] has been developed and applied in order to identify the salient offshore points near the islands, where, the wave energy is concentrated. An artificial unit as shown in Fig. 4.18(b) was adopted for the protection measure. The adopted protection measures for the Vaan Island are shown in Figs. 4.18(c) and 4.18(d). Three different layouts, viz., Case (a) A continuous layer of protection units kept around 250 m from the points on the island periphery which forms a

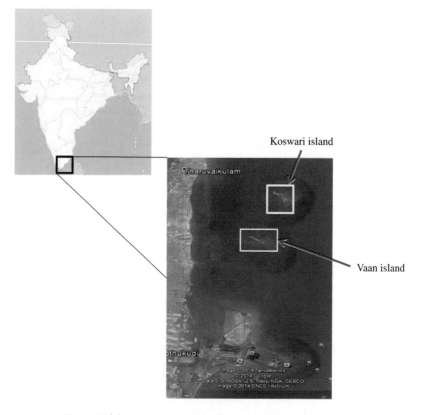

Fig. 4.18(a) Location of the Vaan and Koswari Islands.

continuous curve approximating a near circular curve, Case (b) The same layer but with a gap of 25 m in between individual units and Case (c) An outer layer in addition to the layout as in Case (b) at a distance of 20 m from the same (Fig. 4.18(e)). Based on the wave height distribution, [Case (c)] shown in Fig. 4.18(f), i.e., the two layered protection measure was finalized and implemented. The fabrication of units and its effect on the coral growth are shown in Fig. 4.18(g). The protection measure is found very effective and has also withstood two cyclones.

Fig. 4.18(b) A view of individual reef unit.

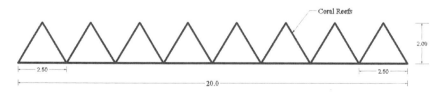

Fig. 4.18(c) Cross-section X-X.

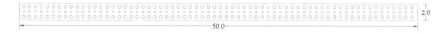

Fig. 4.18(d) Longitudinal section Y-Y.

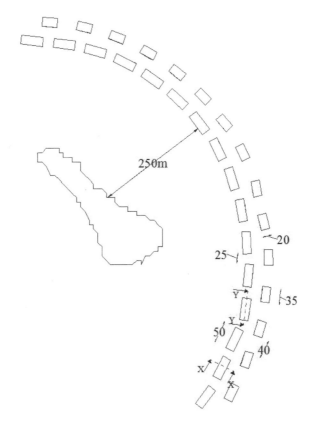

Fig. 4.18(e)   Recommended layout of protection scheme.

4.6.3   Kerala coast

4.6.3.1   General

Kerala Coast dynamic in nature situated along the south west of Indian peninsula is of length of about 590 km, facing the Arabian Sea extends from Latitude 80°15′N to 120°85′N and longitude 740°55′E to 770°05′E several stretches of which are zones of severe erosion.

Until 2006, a single cross-section seawall designed was executed uniformly for the entire coast for protection which had been serving only as temporary solution. The seawalls of uniform cross-sections had been adopted for the entire coast and that was revised over the years the details of which from the 1960s are as described in Fig. 4.19. The seawalls completed using the conventional design suffered severe damage during the subsequent

4. Coastal Erosion and Protection Measures Including Case Studies 129

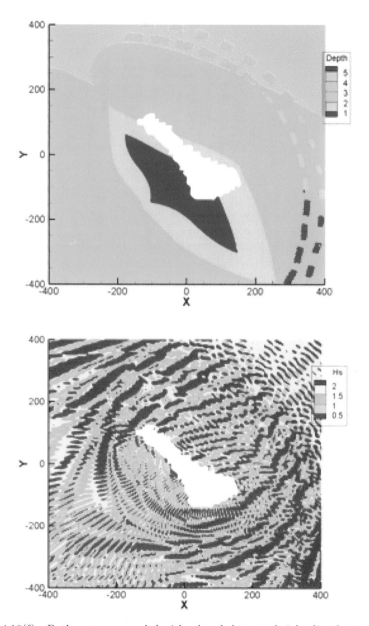

Fig. 4.18(f) Bathymetry around the island and the wave height distributions for the final two layered protection measure.

Pre-casting mould in field | Casting of Continuous units | Accelerating the growth of coarls

Fig. 4.18(g)   Module fabrication and deployment.

erosions. After the tsunami attack of the Kerala coast in December 2004, the authors carried out a study to analyze the damages to the coast during the post tsunami in addition, to the ongoing perennial erosion to arrive at suitable coastal protection measures.

In this regard, the magnitude and direction of sediment movement as well as its mode along these stretches are estimated based on which, suitable coastal protection scheme are drawn and proposed. For locations, where, the littoral drift is less and coastal flooding is severe, it was decided to redesign the seawall cross-section based on the prevailing climate, beach profile changes, wave run-up and sediment characteristics. The cross-sections to be adopted for the study area arrived after a comprehensive study carried out for locations indicated in Fig. 4.20 are provided in Fig. 4.21. The designs proposed for seawalls included geosynthetic fabric filters for underlying layer, Gabion boxes filled with boulders of different sizes for toe and armour. The performance of these structures executed in the three coastal districts are discussed herein.

4.6.3.2   Behaviour of seawalls prior to and after 2008

Panathurakkara coast: (8° 24' 59.8" N; 76° 57' 49.0" E)

From the year 1974 along Panathurakkara coast of Thiruvananthapuram District, Type B, Type C and Type D was tried until 2004, and the seawalls repeatedly failed. After the execution of Type-1 site specific design the seawall has withstood the fury of waves till 2015, with no maintenance after execution. The after and before situation of Panathurakkara coast are shown in Fig. 4.22 and Fig. 4.23.

Maruthadi (8° 54' 25.7" N; 76° 32' 42.2" E)

The major problem had been continuous wave overtopping during monsoon seasons. This area has also got flooded during the ingress of tsunami. The

4. Coastal Erosion and Protection Measures Including Case Studies 131

Crest of seawall +2.74 m, armour layer 110-170 dm^3 with a filter layer of quarry run. (TYPE A, 1964 to 1969)

Crest of seawall +3.35m, armour layer 175-225 dm^3 with a filter layer of quarry run/fascine mattress. (TYPE B, 1969 to 1996)

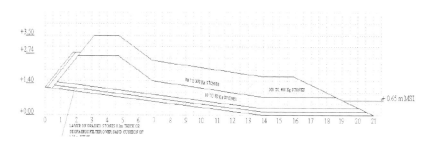

Crest of seawall +3.5 m, armour layer 175-225 dm^3 with a filter layer of geofabric filter with sand cushion over it. (TYPE C, 1996 to 2005)

Crest of seawall +3.90m, armour layer 175-225 dm^3 with a filter layer of quarry run/fascine mattress at selected stretches of the coast (TYPE D, 1971 to 2005)

Fig. 4.19   Designs adapted from 1964 to 2005.

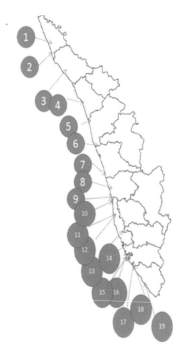

S.no	Name of the Site	Latitude	Longitude
1.	Chervathur	12°12'59.52"N	75° 9'45.40"E
2.	Ettikulam	12° 0'51.55"N	75°12'20.22"E
3.	Kanoor (Azheekal port)	11°56'7.90"N	75°19'53.49"E
4.	Korapuzha	11°23'52.44"N	75°44'34.64"E
5.	Vadakara Kozhikode	11°36'22.85"N	75°35'30.34"E
6.	Ponnani	10°46'3.79"N	75°55'33.24"E
7.	Chettuva	10°32'13.32"N	76° 2'54.95"E
8.	Narayambalam	10° 3'39.73"N	76°12'26.29"E
9.	Cochin	9°55'42.94"N	76°16'4.75"E
10.	Kannamalli	9°52'27.75"N	76°15'49.44"E
11.	Chellanam	9°50'11.11"N	76°16'25.38"E
12.	Chellanam 1	9°48'24.97"N	76°16'38.76"E
13.	Allapey	9°29'44.34"N	76°20'19.68"E
14.	Pallana	9°17'48.03"N	76°23'38.77"E
15.	Azheekal	9° 7'54.54"N	76°27'49.92"E
16.	Jayanthi colony	8°53'31.75"N	76°36'51.82"E
17.	Thangasseri	8°53'6.68"N	76°34'8.68"E
18.	Paravoor	8°48'45.10"N	76°40'2.31"E
19.	Poonthura (Thiruvananthapuram)	8°26'25.15"N	76°56'54.92"E

Fig. 4.20 A few locations of protection measures along the Kerala coast.

overtopping creates scour holes at the interface between the land and the seawall thus leading to its total collapse. In order to reduce the overtopping, gabion boxes of size 1 m×1 m×1 m was proposed as crown wall and executed. This was found to be very effective in the last three years without any damage (Fig. 4.24).

4.6.3.3 Failure of seawalls

As highlighted above, the seawalls have been continuously posing maintenance problems due to its failure. The detailed analysis of failure modes leads to the following salient observations. Due to the low crest elevation, the overtopping of water is inevitable even during high tide. Since there is no provision for drainage facilities on the leeside, the water stagnation poses threat. An inadequate filter layer makes the erosion on the landward side of the crown of the seawall that leads to subsequent subsidence of

4. Coastal Erosion and Protection Measures Including Case Studies   133

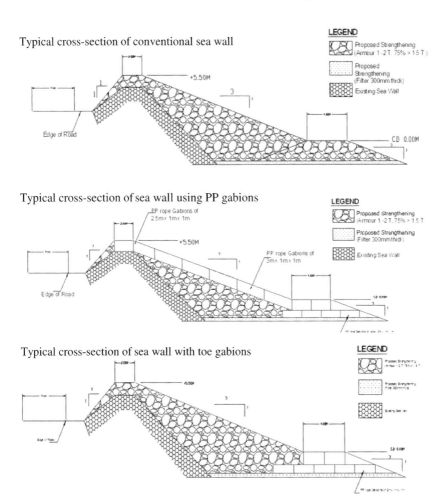

Fig. 4.21   Typical seawall cross-sections proposed and adopted for the Kerala coast.

the structure. Further, the absence of regular maintenance after the storms exposed under layer has exposed the seawall towards its failure. On the sea front, the toe failure exposes the armour stones to induce scour. Further, the design slope of front face would become steepened than the design slope of 1:4. During non-monsoon, alongshore drift dominates in most of the coastal regions of Kerala. This would lead to steepening of the foreshore in front of the seawalls and subsequent toe failure had been occurred.

Fig. 4.22(a) Unprotected mosque at Pathiyarakkara Type B completed and completely sunk in 1988.

Fig. 4.22(b) Mosques protected using Type-1 site specific design in 2009.

4.6.3.4 *Behaviour of groin fields*

About 40 different stretches along the Kerala coast, each of which varying between 0.5 km and 4 km were identified to be dominated by longshore drift. Since the studies carried out and the results derived from implementation the protection measure were found similar the stretch of the coast at Nayarabalam, 10°N and 76°15′E alone is considered for highlighting the studies carried out in arriving at the layout of the groin field and thereafter

4. Coastal Erosion and Protection Measures Including Case Studies   135

Fig. 4.23(a)   Seawall Type C completed and damaged in 2010 (Panathurakkara).

Fig. 4.23(b)   Coast protected using Type-1 site specific design (Panathurakkara).

its role in protecting the coast. Through a rigorous analysis of the available wave data, the longshore current velocity and its distribution across the surf zone as per the methodology of Komar [1969]. The monthly sediment transport has been estimated using the method of Energy Flux [CEM, 2002], the method of Komar [1969] and by integrating the distribution of sediments within the surf zone as suggested by Komar [1969, 1976a]. In view of predominant longshore drift accompanied by the cross-shore sediment movement which was quite visible from the field observation, the groin field was proposed with T-groins. The T-shaped groin can effectively trap the onshore-offshore movement of the sediments. In addition, the long perpendicular arm of the groin can effectively trap the alongshore sediment movement. Following the guidelines laid out in CERC [1984], the layout of the transitional groin field in order to ensure that the downdrift of the

Fig. 4.24 Crown wall using gabion box 1 m×1 m×1 m Maruthadi, Kollam district.

Fig. 4.25(a) Type B seawall damaged, 2008.

extreme groin is affected by erosion was decided. A numerical scheme to solve the one line model using Crank Nicholson implicit finite difference method as suggested by Kraus and Harikai [1983] was understand the shoreline evolution over the years. This was followed by implementation of the groin field followed by its monitoring. The status of the coast prior to 2003 and after 2015 projected in Fig. 4.26 exhibits clearly the success of groin field as coastal protection measure.

4. Coastal Erosion and Protection Measures Including Case Studies 137

Fig. 4.25(b)   Reformation using Type-2 design at Sraikadu.

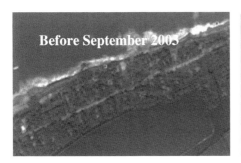

Fig. 4.26 Protective measures along the sea shore against sea erosion at Nayarabalam, Kerala.

4.6.3.5 *Artificial beach nourishment*

Pondicherry (or Puducherry) (11°59'51"N and 79°48'54"E), is a Union Territory town bounded by the southeastern Tamil Nadu state and Bay of Bengal (Fig. 4.27(a)). A pair of breakwaters was constructed in this coastal town to form a harbor. It requires maintenance dredging to be carried out throughout the year at the sand trap due to the prevalence of net littoral drift of 0.6 million m^3 along the east coast. If the maintenance dredging is not carried out regularly, the channel will get closed, and significant erosion takes place on its north, which is the most populous stretch of the city. A total quantity of 0.4 million m^3 of sand was dredged from March to October 2002 from the southern side of the harbour, disposed on its north through a submerged tunnel. The above details are projected in Fig. 4.27(b).

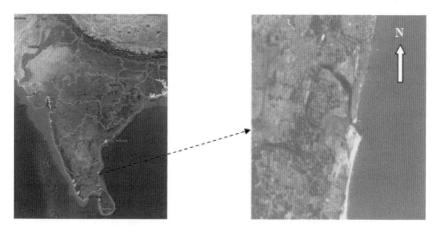

Fig. 4.27(a) Location of Puducherry.

Fig. 4.27(b) Artificial nourishment at Puducherry.

The erosion and accretion is estimated from 1986 to 2002, i.e., over a period of 16 years with the satellite imagery data using GIS software. The rate of erosion is about 4 m per year and the accretion is 6 m per year. The extent of erosion in the northern side is 33.59 hectares compared to the accretion on southern side of 30.71 hectares, see Fig. 4.27(c).

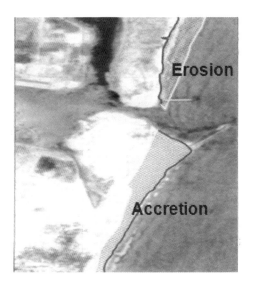

Fig. 4.27(c)   Extent of erosion and accretion.

4.6.3.6   *Geosynthetic products as coastal protection measure*

A stretch of the East Godavari District of Andhra Pradesh coast with several fishing hamlets (17.05°N and 82.31°E), the location of which is shown in Fig. 4.28(a) had been experiencing perennial coastal erosion.

It was decided to protect a portion of this stretch with geosynthetic products. Accordingly, a scheme was drawn out during mid of 2010. The stretch of the coast prior to the implementation of the protection measure is shown in Fig. 4.28(b). The cross-section with a combination of geotextile, geotube and gabions is projected in Fig. 4.28(c). The different stages of construction starting the laying of the geotextile and completing the section are projected in Fig. 4.28(d). It is to be noticed that a few layers of geobags were introduced to serve as a protection for the geotube from being damaged by the gabions. This protection measure has survived several storms, a view of which being exposed to one such event is shown in Fig. 4.28(e).

Fig. 4.28(a)   Location map of Uppada.

Fig. 4.28(b)   Uppada coast before protection.

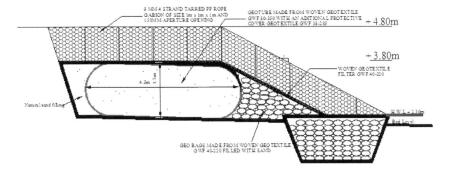

Fig. 4.28(c) Cross-section of the seawall adopted for Uppada coast.

Fig. 4.28(d) Different stages of construction of the coastal protection measures along Uppada.

Fig. 4.28(e) A view of wave action on the seawall during monsoon.

4.7 Assessment of Hard and Soft Measures

Hard Engineering Measures

Structural performance — Coastal defense structures are usually designed for 30- or 50-year design life with minimum or no maintenance at all. The weight of individual units and the thickness of various layers are given as per the provisions mentioned in the shore protection manual. During an extreme storm event the head section of the structures are more susceptible to damage which can be regulated and restored upon consistent maintenance checks.

Overall performance — These structures serves their intended purposes very well. The effect of erosion was successfully overthrown and beach formation was achieved. The heavy mound structures reflect the impact wave energy incident on them and dissipate it leaving minimal impact on the coast, thereby reducing excessive erosion.

A standing example is the groin field at Royapuram. A five decade long problem of erosion due to construction of breakwater for the formation of Chennai port was resorted after careful study and implementation of the groin field.

Table 4.3. Factors governing the adaptation of hard and soft measures.

Factors	Hard measures	Soft measures
Aesthetics	Not appealing	Appealing
Duration for restoration	Immediately effective	Not suitable for immediate enforcement
Cost effectiveness	Huge capital is involved, maintenance is also required in some cases	Capital cost is high for installing slurry pumps; geosynthetic materials are also expensive
Environmental impact	Leaves behind a huge footprint and suitable for almost all climatic and environmental conditions	Leaves behind minimum footprint and their implementation is limited to certain climatic and environmental conditions
Ease of implementation	Speedy and easier	Time consuming and skilled
Design criteria	Shore protection manual and coastal engineering manual	No standard published materials for design of structures using geosynthetics
Design life	30 years to 100 years	Not defined

Soft Engineering Measures and Soft Structures

Structural performance — The design life of structures made up of geosynthetics are comparatively lesser than the rubble mound or concrete structures. There are no standard provisions for design of such geosynthetic units deployed in ocean environment.

Overall performance — Geosystems when combine with other hard measures such as rock armour, gabions etc. perform well. Submerged reefs made up geocontainers, geobags or geotubes which are not exposed to sunlight and anchored properly yield desirable effects.

The conclusions drawn from the foregoing discussion are presented in Table 4.3.

4.8 Tidal Inlets

4.8.1 General

Inlets, as the name implies is an indentation of shoreline leading to small bay usually long and narrow and tide is one of the important reasons for their existence. Some of the inlets are also called as canals or channels. Tidal cycle keeps the inlets open, as the high tide rushes the water into bay; the low tide allows the water back into ocean. The continuous cycle helps the inlet to be open. During a storm surge, it causes a huge amount of water into a bay than what it is used to supply normally during a tidal cycle. When this huge amount of water falls back to the ocean it spills over the barrier island as the risen water has no way of means to retreat and thus it results to an inlet.

For several reasons inlets are quite important but complex. The water inside a bay contains less dissolved salts in it unlike it contains in the ocean water. This kind of water is known as brackish water which falls between fresh and salt water. The survival of several types of marine animals depends on such brackish water inlets. The intensity of salt content in a bay is controlled by the inlets. If an inlet closes, the salt content in the water decreases resulting inhabitable condition for many organisms living in bay.

4.8.2 Tidal flushing

The opening and closing of inlets, and migrate or stabilize of inlets are based on the rate of changes in littoral drift, sediment transport through the inlet, tidal range and flushing, wave climate and also dredging. Variations in an inlet can be seen at different time scales, short term changes can be observed in hours during severe storm events and long term changes in decades or even centuries. As the velocity of current is low at the entrance of inlets, sand particle tend to deposit as shoals. At the seaward boundary outside the inlet, the ebb-tidal deltas occur and deform or withdraw due to the interaction of incoming waves and ebb tides.

The systematic replacement of water in an inlet or bay or estuary as a result of tidal flow is referred to as tidal flushing. The flushing time is defined as the time needed to drain a volume V of water through an outlet A with current velocity U. The flushing time, t_F of an inlet can be defined as the time taken to replace its freshwater volume V_F at the rate of the net flow through the inlet, which is given by the river discharge rate Q. Maximum flushing occurs at spring tides and sometimes during special "monsoon"-periods like in India.

$$t_F = \frac{V_F}{Q} \qquad (4.1)$$

It is typical that the best self-flushing inlets often are placed just "downstream", littoral drift of an obstacle e.g. a headland, or behind a rock reef which, due to its configuration, was helpful in bypassing the drift. Many examples can be found all over the world. The volume of water between the mean high tide to the mean low tide level in an estuary or an inlet is known as tidal prism. The inter-tidal prism volume can be expressed by the relationship:

$$\Omega = HA \tag{4.2}$$

where, H: average tidal range and A: surface area of the basin.

If the tidal prism results in the huge volume of water, meaning that when the tide ebbs, majority of the water is likely to be replaced. Hence, the pollutants or any suspended toxic materials will be flushed away. This occurs in very shallow estuaries. The tidal prism and flushing time of a shallow water inlet is good, whereas, the replacement of water over the entire column of a deeper inlet may not take place.

4.8.3 Stability of an inlet

Evaluation of the stability of tidal entrances on a littoral drift shore must be based on thorough knowledge of tidal hydraulics, wave mechanics, and sedimentary aspects. No absolutely "stable" tidal inlet exists on a littoral drift shore. It is always subjected to changes in its plan form as well as in its cross-sectional area and geometry. Regardless, it maintains its location and cross-section with relatively small changes, due to seasonal changes in wave conditions and to variations in tidal ranges. At tidal inlets on littoral drift shores, the balancing forces are mainly the littoral drift which is carried to the entrance by flood currents for deposits in inner and outer bars, shoals and flats attempting to close the inlet, and ebb tidal and other currents which try to flush these deposits away and maintain the cross-sectional area of the inlet channel.

If an entrance cannot maintain a stable navigation channel by its own flushing capability, then this must be supplemented by artificial means. It is necessary to consider the various methods of improvements that mean control of the inlet including sedimentation. This may be obtained by diversion of flows, dredging of traps etc. or by structural means including measures like training walls, jetties, groins etc.

As per Brunn and Gerritsen [1959], based on a large number of analysis on tidal inlets, classified them as tidal flow bypass and sand bar bypass based on the ratio of annual littoral drift to the discharge from the inlet as below.

- Bar bypassing $[M/Q > 200]$.
- If annual littoral drift, M is very high and tidal discharge, Q is low, the bypassing of sediments would be directly across the inlet. This case leads to the formation of sand bar. Tidal flow bypassing $[M/Q < 7]$.

On the other hand, if the ratio of M/Q is less than 7, the discharge from the inlet flushes away the bar.

The discharge (Q) from an inlet and the littoral drift (M) direction are superposed in Fig. 4.29.

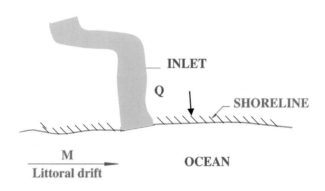

Fig. 4.29   Discharge and littoral drift.

The stability based on the ratio of "Ω/M" was introduced by Bruun and Gerritsen [1960] which is presented in Table 4.4.

Table 4.4.   Stability factor criteria based on "Ω/M".

Stability factor	Navigability
$\Omega/M > 150$	Relatively good condition, good flushing with little bar or no bar formation.
$100 < \Omega/M < 150$	Less satisfactory, and possibly leads to offshore bar formation leading to navigation problem.
$50 < \Omega/M < 100$	May have huge entrance, but there is usually a channel through the high bar, increasing navigation problem.
$20 < \Omega/M < 50$	Inlets are typical "bar-bypassers". Waves break over the bar during storms, and the reason why the inlets "stay alive" at all is often that they during the stormy season like the monsoon get "a shot in the arm" from freshwater flows. For navigation they present "wild cases", unreliable and dangerous.
$\Omega/M < 20$	Unstable entrances, "overflow channels" rather than permanent inlets.

4. Coastal Erosion and Protection Measures Including Case Studies

4.8.4 Stabilisation of tidal inlets

General

A Tidal Inlet may be improved by just dredging, including the installation of sand traps which may be located on the updrift side and where it is most practical and convenient. This is where they are most efficient for trapping of sediments and at the same time present minimum nuisance to navigation. The suitable measure to stabilise an inlet can be identified based on the flow chart given in Fig. 4.30.

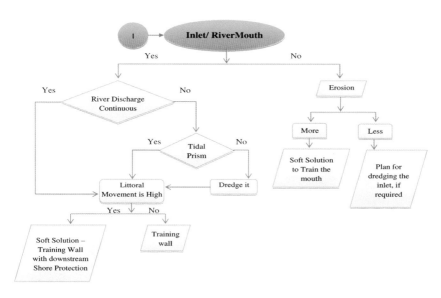

Fig. 4.30   Flow chart for identifying the suitable means to stabilise an inlet.

Equilibrium by structural means

The stability of an inlet can also be maintained by means of installing structural measures, like training walls, jetties, etc. Finally, it may be done by combining structures with sand bypassing schemes. Such schemes may be permanently installed and operated, or they function intermittently by means of traps which are framed, operated or controlled by structures that are shore connected, or offshore breakwaters/submerged reefs including improved natural reefs.

Structures for improvements include:

(a) Jetties or training walls possibly and/or with spurs if the banks of the inlet need to be stabilised
(b) Breakwaters built parallel to shore
(c) Weirs in breakwaters with traps inside for periodic dredging
(d) Dredging of navigation channel

Jetties are the most common adopted structural measure. They intercept the free passage of the littoral drift and prevent or delay the formation of sand bars in the channel. The efficiency depends upon the length of the jetties in relation to bottom slope, wave exposure and the drift of sediments, mainly sand, caused by the waves. Jetties may be single or double. Generally, it is true that the single jetty is not a good protective measure. Its effect is beneficial updrift, but downstream it creates two problems:

(a) Erosion of the downdrift shore.
(b) Deposition in a leeside eddy, which encroaches upon the navigation channel when expanding updrift.

It is therefore not advisable to build single jetty/groin unless a special natural entrance configuration invites one to do so, but there are only a very few cases of nature controlling. The lengths of the pair of jetties, (Fig. 4.31(a)) are the most important parameter, the length of which on the updrift side is usually longer and has to certainly pierce the surf width for it to trap the sediments effectively from silting the mouth of the inlet. The length is often determined based on the normal undisturbed distribution of drift in the profile or offshore area, but this distribution will be often considerably altered by the presence of the jetties. Sometimes updrift side jetties are provided with small "spurs" that means jetties built perpendicular to the main jetty to trap sand. Double training walls on the updrift side also constitute a littoral drift barrier. One of the methods adopted for pumping sand across an approach channel is by means of floating pipelines mounted on pontoons. It would be necessary to detach the pipeline frequently for the movement of boats in the channel and as such, during periods of heavy traffic, bypassing will be seriously restricted. Under these circumstances, pipelines buried below the bed of the channel could be adopted for the uninterrupted bypassing of drift across the inlet. However, it is obvious that this pipeline would be considerably more expensive and heavier than the floating pipeline as this has to be withstand the pressure of the water column above.

4. Coastal Erosion and Protection Measures Including Case Studies   149

The above said concepts along with the three most popular protection measures in the event of erosion on the downdrift side are projected in Fig. 4.31(b).

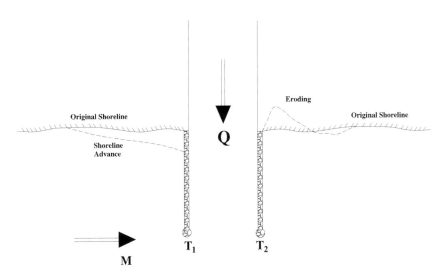

Fig. 4.31(a)   A pair of training walls for maintaining the stability of an inlet.

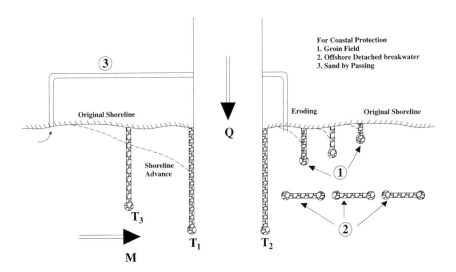

Fig. 4.31(b)   Training walls and associated protection measures.

It may be "partial", cutting off part of the drift, or perhaps "complete", that means all the longshore drift is cut off. In most cases only part of the drift is cut off, and a certain percentage of drift bypasses at the same time as accumulation takes place updrift. If the training walls lacks the required velocity to flush the bar formation, narrowing the training walls at the entrance enhances the velocity to flush the bar formation to attain equilibrium.

Sometimes improvements by dredging and by structures are combined in that jetties or training walls are built to include bypassing arrangements of permanent or intermittent nature.

Equilibrium by means of bypassing

Seen in larger perspective, it may be said that bypassing by sand pumps aids two purposes: 1) protection of navigation channels against deposits by longshore littoral drift materials; 2) protection of downdrift side beaches against starvation caused by the littoral drift barrier. Various methods to bypass the sediments are discussed below.

- Sand pump on a Trestle updrift of the inlet
- Sand pump at the tip of the updrift Groin [on a short trestle]
- Sand pump on a Trestle downdrift of the inlet

Sand pumps are normally employed to re-establish the littoral drift obstructed by a groin updrift of a tidal inlet, which is maintained artificially so as to avoid erosion at the downdrift of an inlet. For this purpose, the sand pump is normally located on a jetty or mounted on a trestle working over a sand trap updrift of the inlet Fig. 4.32(a). However, the disadvantages of the sand pumps are lack of adequate mobility, their inability to undertake dredging during high seas, when, a large quantity of drift is expected to take place along the coast. In the event of a breakdown of the pump, during a monsoon season, a shallowing of the updrift region takes place resulting in a severe shoaling of the channel.

This could result in the shifting of the region of maximum drift away from the location of the sand pump so that dredging during the subsequent monsoon would be rendered rather inefficient.

In rare cases, the sand pumps can also be deployed at the downdrift of the inlet over a trestle. In such a case, the approach channel itself acts as a sand trap in addition to the trap at the updrift. Figure 4.32(b) depicts the location of sand pump at the downdrift of inlet.

In order to avoid the drawbacks inherent in fixed bypassing plants, a floating dredger could be adopted for bypassing sand across the inlet more conveniently. For this purpose, a sand trap is provided updrift of the

4. Coastal Erosion and Protection Measures Including Case Studies 151

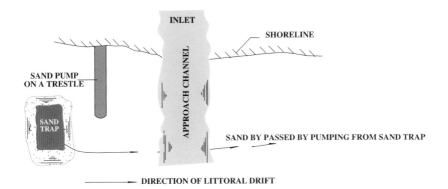

Fig. 4.32(a) Sand pump on a Trestle updrift of the inlet.

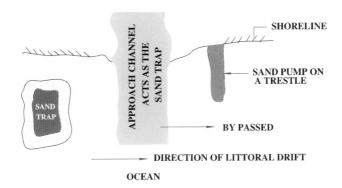

Fig. 4.32(b) Sand pump on a Trestle downdrift of the inlet.

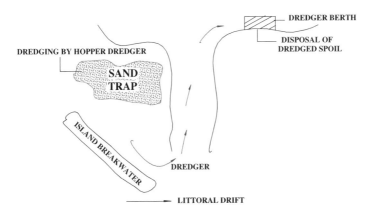

Fig. 4.32(c) Floating dredger to bypass the sand and approach channel protected by breakwater.

approach channel and protected generally by an island breakwater. And the material dredged at the trap is pumped directly downdrift by the pipeline of the dredger.

Alternatively, for large inlets, the dredging is carried out by a hopper dredger and the materials are taken to a dredger berth located downdrift of the inlet from where the materials could be directly pumped, thereby avoiding the pipeline from crossing the channel [Fig. 4.32(c)].

4.8.5 Crater — Sink sand transfer system

This technique originally suggested by Inman and Harris [1970] consists of a hydraulic jet assembly operating from the bottom of the sand crater.

The sand transfer system involves of mouth section, a jet pump, transfer and delivery pumping of slurry sand mixture facility. As the mouth section is situated at the lowest point, the entire crater acts as gravity fed for all the materials in the sink. Figure 4.33(a) shows the schematic plan view of entrance channel with crater sink sand transfer system. Figure 4.33(b) shows the cross-section of crater sink sand transfer system. The transported sand are left at the sides of the crater for suction which slides to the mouth and available for transfer. This serves as a dual purpose of collecting and accumulating of sand.

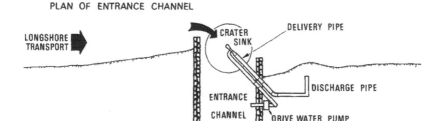

Fig. 4.33(a) Schematic plan view of crater-sink sand transfer system when used to bypass sand across the entrance to a harbour or lagoon.

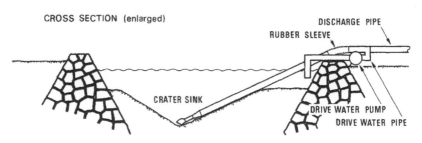

Fig. 4.33(b) Schematic cross-section of crater-sink sand transfer system.

4.9 Case Studies on Tidal Inlets

The sand bar formation and chocking of river mouths is one of the most commonly occurring phenomena along the coast of India. The river mouths along Cheruvathur, and Chettuvai along the coast of Kerala situated on the South west coast of India were identified to undergo severe choking and predominant sand bar formation preventing the ease of flow of river stream water into the ocean.

Cheruvathur creek is located on the south west coastline, (12°11′59″N and 75°7′12″E). The site dominated by net southerly drift of sediment transport, resulted in sand bar formation at the confluence of the river and Arabian Sea. This necessitated a detailed study on suitable remedial to

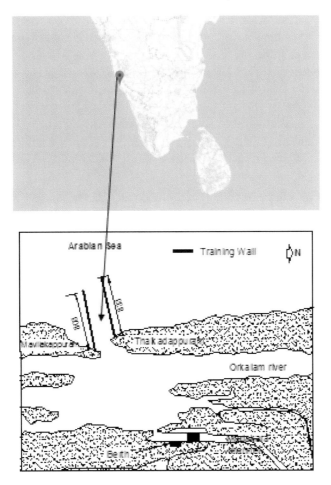

Fig. 4.34 Location of Cheruvathur, Kerla.

prevent the sand bar formation and to establish a smooth flow of water from the river into the ocean. The location map of the proposed site is shown in Fig. 4.34.

Based on the field studies of hydrographic and oceanographic data collection, nearshore data, satellite imageries and shoreline changes it was decided to achieve equilibrium by structural means. The proposed measure should also beneficially satisfy the needs of the fishermen community, where, it could also alleviate maneuvering problems of fishing boats.

The canal mouth on the southern side of the creek (near the creek mouth) was found to be fully blocked by the formation of sand bars. This is clearly due to the predominant southerly drift. In addition, perennial sand deposition on the northern banks near the mouth was noticed. This may be due to the northerly sediment drift in the ocean. Based on analyzing these conditions, the alignment of training walls had been decided. The length of the northern training wall is 0.833 km extending up to a water depth of 5 m and the southern training wall is 0.803 km long extending up to a water depth of 4 m. One of the training walls was inclined in order to achieve adequate discharge velocity to flush the sand bar also to bypass the littoral drift. The variations of the shoreline adjoining the pair of training walls near the mouth of Cheruvathur creek prior and after the implementation of the solution are projected in Figs. 4.35(a), 4.35(b) and 4.35(c).

Fig. 4.35(a)   Cheruvathur creek, Kerala (2005).

4. Coastal Erosion and Protection Measures Including Case Studies 155

Fig. 4.35(b) Shoreline changes of Cheruvathur creek, Kerala (2010).

Fig. 4.35(c) Shoreline changes of Cheruvathur creek, Kerala (2015).

Fig. 4.36 Google earth imagery of Chettuvai.

In order to ascertain the flushing away of the sediments that is likely to be deposited near the mouth, the fall velocity, v_f was calculated for the mean grain size, D of 0.25 mm based on the relations discussed in Sec. 3.6. In this way, the fall velocity works out to be 0.1 m/s. This is extremely small compared to the velocity in the river Cheruvathur.

4. Coastal Erosion and Protection Measures Including Case Studies 157

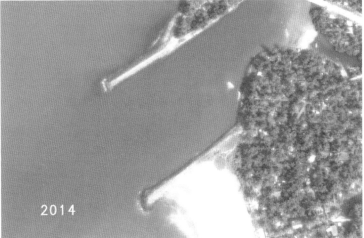

Fig. 4.37   Google earth imagery of Kallai.

Similar problems of river mouth choking and sand bar formation was also observed at various sites in India. The training of river mouths was evaluated as a viable solution and based on the field studies the alignment of training walls was distinguished according to the sites. The imageries on the results from the other sites, namely Chettuvai (10°30′29.72″N, 76°2′20.33″E) and Kallai (11°13′37.77″N, 75°46′44.89″E) are projected in Figs. 4.36 and 4.37, respectively.

References

Berkhoff, J. W. (1972). Computation of combined refraction-diffraction, *13th Int. Conf. Coastal Eng.*, Vancouver.

Bruun, P. and Gerritsen, F. (1959). Natural bypassing of sand at coastal inlets, *J. Waterways and Harbours Div.*, 85, 75–107.

Bruun, P. and Gerritsen, F. (1960). *Stability of Coastal Inlets*, North-Holland, Amsterdam.

Coastal Engineering Manual, CEM (2006). U.S. Army Corps of Engineers, Coastal Engineering Research Centre, Vicksburg, Mississippi.

Clark, J. L. (1995). *Coastal Zone Management Handbook*, Lewis Publishers, Washington, D.C.

French, P. W. (2001) *Coastal Defences, Processes, Problems and Solutions*, Routledge, New York.

Hanson, H. and Kraus, N. C. (1990). Shoreline response to a single transmissive detached breakwater, *Coastal Engineering Proceedings*, 1.

Inman, D. L. and Harris, R. W. (1970). Crater-Sink Sand Transfer System, Chapter 58, *Proc. 12th Conf. Coastal Eng.*, Washington, D.C., New York.

Janardanan, K. and Sundar, V. (1997). Effect of uncertainties in wave characteristics on shoreline evolution, *J. Coast. Res.*, 13(1), 88–95.

Komar, P. D. (1969). The Longshore Transport of Sand on Beaches, Ph.D. Thesis, Scripps Institution of Oceanography, University of California, San Diego.

Komar, P. D. (1976a). Beach process and sedimentation, Prentice-Hall, 429 pp.

Kraus, N. C. and Harikai, S. (1983). Numerical model of the shoreline change at Oarai beach, *Coastal Engineering*, 7(1), 1–28.

Li, B. (1994). A generalised conjugate gradient model for mild slope equation, *Coastal Engineering*, 23, 215–225.

Pilarczyk, K. W. (2000). *Geo-synthetics and Geo-systems in Hydraulic and Coastal Engineering*, A. A. Balkema, Rotterdam.

Sundar, V. (2005). Behaviour of shoreline between groin field and its effect on the tsunami propagation, *Fifth International Symposium on Wave Measurement and Analysis*, Madrid, 3–7 July, Spain, paper No: 323.

U.S. Army Corps of Engineers. (2002). Coastal Engineering Manual (CEM), Engineer Manual 1110-2-1100, U.S. Army Corps of Engineers, Washington, D.C. (6 volumes).

Verhagen, H. J. (1992). Method for artificial beach nourishment, *Proc. 23rd Int. Conf. Coastal Engineering*, pp. 2474–2485.

Chapter 5

Rubble Mound Structures

5.1 Introduction

Rubble mound structures are constructed for the construction seawalls and groins for coastal protection, revetments for bank protection, and training walls for prevention of sand bar formation in a dominant littoral zone or for navigational channels, spurs for training of rivers, breakwaters for the formation harbour and coastal protection through detached offshore or as submerged breakwaters. The design of a rubble mound structure is quite difficult compared to solid face structures as the interaction waves with such structures is complex and are dictated by empirical formulae arrived through physical modelling. Waves usually break over rubble mound structures and are exposed to non-stationary velocity and acceleration fields in particular during their run-up and rundown. The protection against waves offered by these structures mainly depend on its permeability and slope. Flatter the slope, gradual is the dissipation of incident wave energy and individual armour blocks that are needed for the structure to be stable will be. Steeper slopes require larger size armour blocks for its stability. The wave characteristics, tidal range, extreme events like storm surge control the design of rubble mound structures. Optimizing a rubble mound structure that satisfy all the requirements can only be achieved through comprehensive physical model tests. The breakwaters, jetties, revetments, seawalls, and spurs are mostly formed with a core of natural small size natural rocks, sand filled in tubes (geo-tubes), or other suitable material over which a layer of natural rocks, the weight of which is a function of the prevailing wave climate are laid. For breakwaters, particularly in deeper locations of high waves, an additional layer (top or primary layer) of artificial armour blocks of concrete are adopted. The rubble mound structures adopted for coastal protection or for formation of harbours are exposed to waves propagating over different water levels, thus necessitating specific designs or many times it may be a conventional one as per the site conditions and the purpose for which it is constructed. They are either designed as wave overtopping or non-over topping type. In the case of an overtopping types, a crown wall over the top of the breakwater or seawall is positioned at the crest elevation of the structure. Although empirical relationship exist to determine the weight of the individual armour block, its stability is usually ascertained through physical model tests. At certain locations, berms on the sea side are also employed. The physical modeling is carried out also to investigate the extent of wave run-up over the structure, overtopping, reflection, absorption and transmission of waves. Such tests are carried out in flumes. However, the layout of breakwaters, like its length, height and orientation with respect to wave direction for the formation of artificial harbours are finalized through three-dimensional wave tanks, in which the interaction of coastal structures as well as their stability under oblique waves are determined. In both two and three D tests, a larger scale factor is adopted to reduce the errors in matching the results closer to that could be in the field.

Although, this chapter is devoted to rubble mound structures, it is equally important to have a broader idea on breakwaters.

5.2 Types of Breakwaters

Breakwaters are broadly classified as mound type, vertical type and composite type. A mound type breakwater does not produce substantial wave reflection, attributed to the incident wave energy breaking on a slope of rubble or concrete armour layer and they tend to dissipate most of the incoming wave energy. Vertical breakwaters can either stand on sea bed or can be built on artificial rubble mound foundations termed as composite breakwaters. On the other hand, vertical breakwaters reflect the incident wave energy as the breakwater is impervious/non-porous. Composite breakwaters are employed in locations, where, the tidal fluctuations are significant; so that they function as mound type during the low tide and as vertical type during the high tide. Based on the location it serves, breakwaters can further be classified as detached, headland, nearshore and shore connected breakwaters. They may emerge, submerge the MSL or may be floating.

- Type S called sloping breakwater because its seaward face has a slope of 1:1 or flatter. These breakwaters basically are of rubble mound.
- Type V called vertical breakwater, because it has a vertical or nearly vertical seaward face.

- Type C called composite breakwater consisting of a vertical breakwater, placed on a foundation with slopes on both sides, in which case, the waves are reflected at high water, but break at low water.

The large majority of sloping breakwaters consists of rock and/or concrete blocks, the widely used name is rubble-mound breakwater. In deeper waters, they usually have a concrete superstructure (limits the overtopping at minimum cost, access for traffic, other needs), in shallow water they rarely have a real superstructure.

Upright wall breakwater: An upright wall with block masonry was initially popular, in which, many different methods were applied to strengthen the interlocking between the blocks. The wall extends over the full depth of water, or most of it. The main disadvantage is the large overturning moment due to wave reflection resulting in sudden failure in case the design height is exceeded (Fig. 5.1(a)). Its foot print on over the seabed is less compared to the S-type.

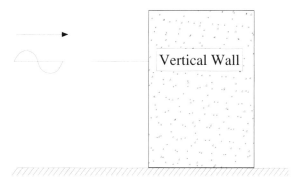

Fig. 5.1(a)   Basic vertical breakwater.

Caisson breakwater as wave energy convertor

The seaside of vertical caisson breakwaters can be modified to serve as a wave energy convertor (WEC) and its leeside can still be used for berthing of vessels. The oscillating water column device has a chamber with its one end open to the sea and the other end vented to an air turbine. A propagating wave crest acts as a piston and compresses the air inside the chamber thus, forcing it out through the turbine. When the wave trough is present in the chamber, air is sucked back through the turbine. Hence, a turbine that can rotate in the same direction irrespective of the direction of air flow is necessary to drive an alternator in order to *generate electricity*, on both cycles. Such a turbine is called "Wells turbine". The working principle of OWC and its integration with berthing facility is illustrated in Fig. 5.1(b). Sundar *et al.* [2010] presented a comprehensive review on novel approaches that can utilize the wave energy converters as part of coastal defence systems and breakwaters for the formation of harbours along with a brief review of OWC around the globe.

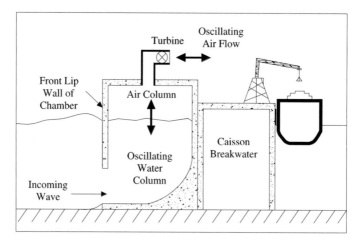

Fig. 5.1(b) Basic vertical breakwater.

Apart from the above stated types, there are special types of breakwaters. A type of breakwater introduced as a special type would become a conventional over a period after it becomes quite popular.

Semi-circular breakwaters

It is a semi-circular shaped hollow caisson with fully perforated, founded on a rubble mound. It is cast as different elements and made of prestressed concrete. Since the caisson is hollow, its weight and the materials to be used are significantly less. The stability against sliding for semi-circular breakwater is good, since, the horizontal component of the wave force is smaller as compared to the vertical component; in addition, the vertical component is applied downward the curved wall. It enhances the scenery compared to the conventional rubble mound breakwaters if, semi-circular breakwater is used as surface piercing one. According to Sasajima et al. [1994] this type of breakwater is broadly classified as shown in Fig. 5.2. These are impermeable, seaside wall perforated, fully perforated and leeside wall perforated. Fully permeable type offers maximum wave dissipation. This type of breakwater is constructed in Miyazaki Port in 1993.

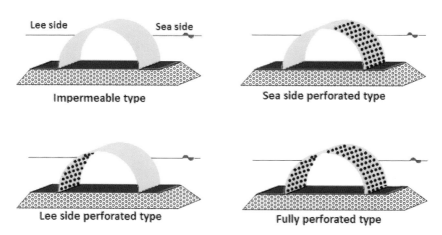

Fig. 5.2 Semi-circular fully perforated breakwater.

Herein, only S-type breakwaters are discussed although an overview on other types are briefly discussed.

5.3 Criteria for Breakwater Selection

Amongst the numerous types of breakwater options available for construction at a given site, the significant wave height, H_s and water depth, d are the most important primary criteria for breakwater type selection.

Type S

If the topography has good rock and large blocks then the Type S breakwater with or without superstructure or Type C breakwater shall be considered under following conditions:

$H_s < 3$ m: Type S without superstructure
$H_s > 3$ m, $d < 20$ m: Type S with superstructure
3 m $< H_s < 6$ m, $d > 20$ m: Type S with superstructure
$H_s > 6$ m, $d > 20$ m: Type S with superstructure or Type C

When good rock is available but the blocks are not large enough then its best to go with Type S but the armour should consist of concrete blocks in order to have stability. The Type S breakwater can be used when there is a short breakwater or small vessels like fishing boats, pleasure crafts are at harbour entrance.

Type V

When there is a reasonably good foundation but lack of good rocks, then we can choose either Type V if $d < 15$ m or Type C if $d > 15$ m.

The caisson Type V breakwater can be preferred for the following conditions such as, bad weather all year or rapid change in wave conditions, bad weather for several months and also when the offshore breakwater is with berth on lee side.

Type V is more suitable for condition which require enhanced sediment bypass at the harbour.

Based on labour availability, if there is cheap skilled labour then caisson breakwater can be preferred for easy installation whereas if costly labour is available with low cost equipment then rubble-mound is the best option.

Typical view of the breakwater fronted with rubble mound of Chennai Harbour, India during the attack of a storm in 2003 is projected in Fig. 5.3.

Fig. 5.3 Violent wave attack on the composite breakwater of Chennai port in late 2003.

Rubble mound or S-type breakwaters

Rubble mound breakwaters are frequently used structures. It consists of armour layer, under layer, core and filter layer. It is a structure, built up of core of quarry run rock overlain by one or two layers of large rocks. It is easier to construct and damage normally occurs due to poor interlocking capacity between individual blocks. Where large size natural rocks are not available, artificial armour blocks are adopted, which will be discussed later. Armour layer stability can be increased by using shape designed concrete blocks, referred to as "artificial armour blocks" while, wave overtopping can be reduced using a super-structure or crown wall over the top of the mound (Type S with Superstructure). It is constructed by a heterogeneous assemblage of natural rubble or undressed stone.

When water depths are large rubble mound breakwater may be uneconomical in view of huge volume of rocks required. It is unsuitable at locations when space is a constraint. These S-type of breakwaters dissipate the incident wave energy by forcing them to break on a slope and thus do not produce appreciable reflection. Energy lost due to friction, percolation through the porous medium and due to partial reflection. Due to their armour, they are more durable and can be easily adapted to irregular bathymetry. The stone and gravel accommodates settlement and thus being flexible too. The gravel and sand cannot be laid on sand and hence requires under layer if built on sand and also these are not suitable for soft ground. The armour unit can be either quarry stones or concrete units depends on the availability. The placement of armour units can be both random or regular arrangements which depends on the type of armour units and the wave climate it exposed to. Typical cross-sections of conventional rubble mound and S-type with superstructure are projected in Figs. 5.4 and 5.5, respectively. The seaward slope is usually flatter (1:2 or 1:2.5), whereas, the harbour side slope is steeper (1:1.5 or 1:2). The slope of the head of the breakwater is still flatter (1:3 to 1:5).

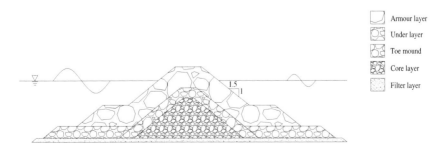

Fig. 5.4 Typical cross-section of S-type rubble mound breakwater.

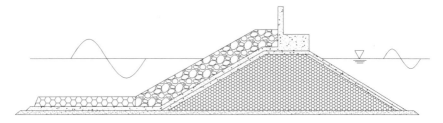

Fig. 5.5 Typical cross-section of S-type with Superstructure.

Reshaping or berm breakwaters utilize the basic concept of establishing an equilibrium between the slope of the rubble mound and wave action i.e. the rubble mound forms a S-shape slope to stabilize itself against wave action. Typical cross-section of a berm breakwater is shown in Fig. 5.6.

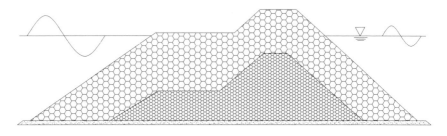

Fig. 5.6 Typical cross-section of berm breakwater.

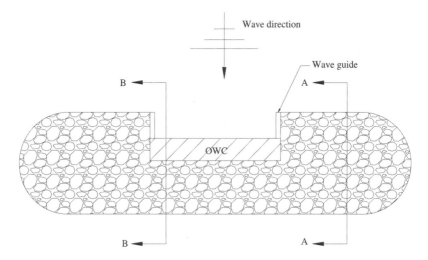

Fig. 5.7(a) Plan view of OWC integrated with offshore detached breakwater.

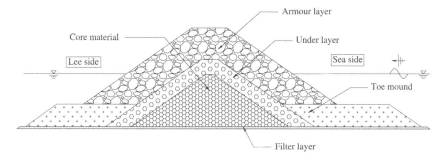

Fig. 5.7(b) Cross-section of rubble mound breakwater at A-A.

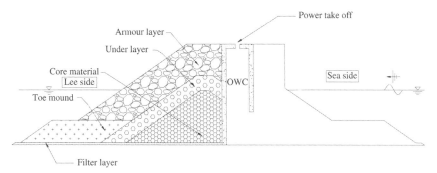

Fig. 5.7(c) Cross-section of OWC integrated with rubble mound breakwater at B-B.

An OWC can also be integrated with an S-type breakwater similar to that with caisson breakwater as discussed earlier. This could be with one or more numbers of OWC integrated with an S-type breakwater for the formation of a harbour or, to serve as part of detached breakwaters to serve as coastal protection measure. Thus, the costs will be shared and the benefits are more than just coastal protection or energy from ocean waves. A typical concept of incorporating OWC with S-type breakwater for this purpose is shown in Figs. 5.7(a)–5.7(c). The performance characteristics of an array OWC integrated with offshore detached S-type breakwaters subjected to different angle of wave incidence in regular wave field have been assessed through a well-controlled comprehensive experimental program by John Ashlin *et al.* [2016] and the results are found to be quite encouraging. Typical view of OWC integrated rubble mound and semi-circular detached breakwater that could be adopted for islands are shown in Figs. 5.7(d) and 5.7(e), respectively. A typical deep water berthing facility with OWC is shown in Fig. 5.7(f).

168 Coastal Engineering: Theory and Practice

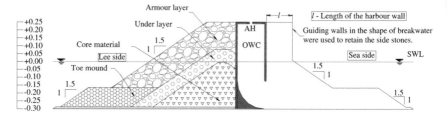

Fig. 5.7(d) Typical rubble breakwater with OWC and its cross-sectional view at A-A.

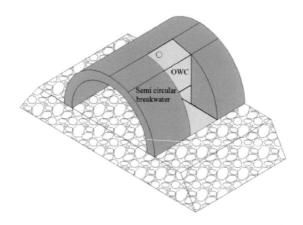

Fig. 5.7(e) Typical semi-circular breakwater with OWC.

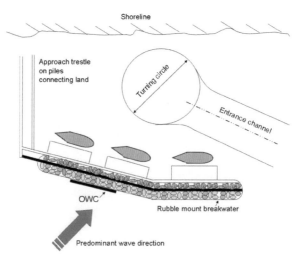

Fig. 5.7(f) Typical deeper water direct berthing facility formed by breakwater with OWC.

This encompasses an approach trestle (an open jetty supported on piles), a breakwater (may be of caisson type or rubble mound type). Any of the foresaid OWC system can easily be integrated into breakwater or even with the approach trestle. This however needs a careful and detailed investigation.

5.4 Design Principles of Rubble Mound Structures

The design of rubble mound breakwater involves establishing the unit weight of armour stones which is primary requirement. The weight of armour stones for the primary layer is based on the Hudson's formula given by

$$W = \frac{\gamma_s \times H^3}{K_D \times (S_r - 1)^3 \cot \alpha} \tag{5.1}$$

γ_s: Unit weight of angular quarry stones or artificial armour blocks = 2.65 t/m^3 and 2.76 t/m^3 for quarry stones.

H: Design wave height at toe of breakwater

K_D: Stability coefficient = 3.5 for roughly angular randomly placed quarry stones

$\quad\quad\quad\quad\quad$ = 6.0 for tetrapods.

$$\delta_r: \frac{\gamma_s}{\gamma_f} = \frac{\text{unit wt of material}}{\text{unit wt of water}} \simeq \frac{2.65}{1.03} \simeq 2.57$$

α: Slope of the breakwater (for a slope 1:2, Cot α = 2)

The dimensionless stability coefficient, K_D in Hudson formula accounts for all variables other than structure slope, wave height and specific gravity of water at the site.

These variables include:

1. Shape of armour units
2. Number of units comprising the thickness of armour layer
3. Manner of placing armour units
4. Surface roughness and sharpness of edges of armour units (degree of interlocking of armour units)
5. Type of wave attacking structure (breaking or non-breaking)
6. Part of structure (trunk or head)
7. Angle of incidence of wave attack

The weight of the individual armour unit is proportional to H^3 which warrants careful evaluation of H. For deeper waters wherein, the wave height expected to be more or at locations wherein natural rocks are scanty, artificial concrete armour blocks are considered. The K_D factor depends on the interlocking capacity of the blocks, and efforts have been to increase it by varying the shape of the blocks.

5.5 Concrete Armour Layer Units

5.5.1 *General*

The most commonly adopted breakwaters are the S-type, in which the decision on the selection of artificial armour blocks particularly at locations, wherein, natural rocks are scanty is vital. Hence this section focuses on the said aspect. The CAUs — concrete armour units were developed in order to address this limitation to natural rocks. Though, the CAUs are heavy, it is also designed for dissipating the wave energy by making it porous.

Many types of armour unit shapes are available and each has its own engineering performance characteristics. Armour blocks too have various qualities that contribute to their stability. Some of the important ones are, weight, interlocking, inter-block friction, hydraulic drag, permeability of assembly etc. Typical armour units include the Dolos, Tribar, Tetrapod, Accropode and block etc. The default armour unit is stone for both economic and aesthetic reasons. CAUs are used when the costs associated with quarrying, transporting, and placing stone armour large enough to be hydrodynamically stable exceed those of CAUs. A general idea about the pros and cons of different armour units developed in the past decade are discussed below.

The widely adopted forms of CAUs available are shown in Fig. 5.8.

5.5.2 *Randomly placed armour units — stability factors weight and interlocking*

The randomly placed armour units have been in practice from 1950 and have developed over the period for various advantages. These can be classified into three generation based on the shape, stability and year as follows.

First generation armour units

The shape of first generation armour blocks is quite simple in its shape and the structure is usually formed by two layers or more in exceptional cases and is placed in random. The stability of the block relies on its gravity and

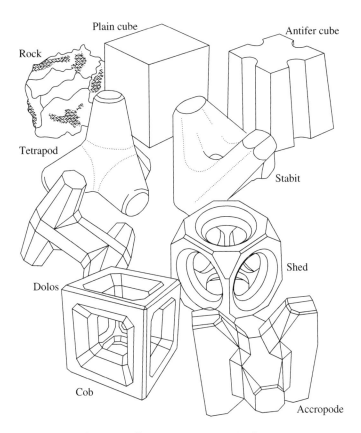

Fig. 5.8   Concrete armour unit shapes.

interlocking capability with adjoining blocks. There has been a continuous effort by researchers in reshaping the blocks in order to enhance its interlocking capacity as well as increase its K_D.

Second generation armour units

The second generation armour units can be classified into two types based on their shape profile,

(i) Simple shape: Similar to the first generation, these simple shape second generation has the stability factor as weight and to some extend interlocking. The placement pattern is mostly random and in two layers. Typical examples of CAUs under this category are Tetrapod, Tribar, Tripod, etc.

Tetrapod

These are tetrahedral blocks used as armour for the breakwaters. Its shape allows it to dissipate the incoming wave, thus allowing the water to flow around. It reduces the displacement of rubble mound by allowing a random distribution of tetrapods to interlock. The tetrapods tend to get dislodge by the crashing wave and hence they are often numbered to monitor any displacement.

(ii) Complex shape: Because of its complex shape, the interlocking of armour blocks governs the stability factor. Distinctive type of this generation armour units are stabit and Dolos.

Dolos

These are made from un-reinforced concrete, poured into a steel mound. In the absence of reinforcement, steel fibers are mixed to the concrete for higher stability. These are limited to be installed near the coast for its heavy weight. These are used to protect harbour walls, breakwater and shorelines. They can trap sea-sand thus preventing erosions. They are highly stable and do not dislodge easily. They dissipate the waves by deflecting the wave energy. The stability coefficient, K_D which depends on the number of layers to be adopted, its slope as well as the nature of waves (breaking or non-breaking) is derived through physical model tests by subjecting the cross section to the action of 500 to 100 waves. The K_D for the different CAUs are projected in Table 5.1.

5.6 Kolos

5.6.1 *General*

Kolos, a concrete armour module is designed and developed from Dolos. It is regarded as a two elongated concrete members connected on opposite sides by a short central shank. The longitudinal axes of outer elongated members are parallel and linked with a central shank of which the longitudinal axis is normal to the outer members. The cross-section of outer members are octagonal with decrease in area towards the end. An isometric view of a Kolos is illustrated in Fig. 5.9. Through a comprehensive physical model testing program, the authors have found the range of stability coefficient for KOLOS for the trunk and head portion were found to be in the range, 23 to 32 and 3.5 to 8 for the trunk head portions, respectively. The above said details are discussed under Sec. 5.7.

5. Rubble Mound Structures 173

Table 5.1. K_D values for use in determining armour unit weight.

Armour units	n^a	Placement	No — Damage Criteria and Minor Overtopping					Slope
			Structure Trunk		Structure Head		
			K_D^b		K_D	
			Breaking Wave	Non-Breaking Wave	Breaking Wave	Non-Breaking Wave	Cot θ
Quarrystone							
Smooth rounded	2	Random	1.2^c	2.4	1.1	1.9	1.5 to 3.0
Smooth rounded	>3	Random	1.6	3.2	1.4	2.3	d
Rough angular	1	Random^b	b	2.9	b	2.3	d
					1.9	3.2	1.5
Rough angular	2	Random	2.0	4.0	1.6	2.8	2.0
					1.3	2.3	3.0
Rough angular	>3	Random	2.2	4.5	2.1	4.2	d
Rough angular	2	Special^e	5.8	7.0	5.3	6.4	d
Parallelepiped^f	2	Special^c	7.0–20.0	8.5–24.0	—	—	
Tetrapod and					5.0	6.0	1.5
Quadripod	2	Random	7.0	8.0	4.5	5.5	2.0
					3.5	4.0	3.0
Tribar				8.3	9.0	1.5	
	2	Random	9.0	10.0	7.8	8.5	2.0
				6.0	6.5	3.0	
Dolos	2	Random	15.8^g	31.8^g	8.0	16.0	2.0^h
				7.0	14.0	3.0	

(Continued)

Table 5.1. (Continued)

			No — Damage Criteria and Minor Overtopping				
			Structure Trunk		Structure Head		Slope
			K_D^b		K_D		
Armour units	n^a	Placement	Breaking Wave	Non-Breaking Wave	Breaking Wave	Non-Breaking Wave	Cot θ
Modified cube	2	Random	6.5	7.5	—	5.0	d
Hexapod	2	Random	8.0	9.5	5.0	7.0	d
Toskane	2	Random	11.0	22.0	—	—	d
Tribar	1	Uniform	12.0	15.0	7.5	9.5	d
Quarry stone (K_{RR}) Graded angular	—	Random	2.2	2.5	—	—	

[a]n is the number of units comprising the thickness of the armour layer.
[b]The use of single layer of quarrystone armour units is not recommended for structures subject to breaking waves, and only under special conditions for structures subjects to non-breaking waves. When it is used, the stone should be carefully placed.
[c]Caution: those K_D values shown in italics are unsupported by test results and are only provided for preliminary design purposes.
[d]Until more information is available on the variation of K_D values with slope, the use of K_D should be limited to slopes ranging from 1:1.5 to 1:3; some armour units tested on structure head indicate a K_D-slope dependence.
[e]Special placement with long axis of stone placed perpendicular structure face.
[f]Parallelepiped-shaped stone: long slab-like stone with the long dimension about three times the shortest dimension [Markle and Davidson, 1979].
[g]Refers to no-damage criteria (<5% displacement, rocking etc.); if no rocking (<2%) is desired, reduce K_D by 50% [Zwamborn and Van Niekerk, 1982].
[h]Stability of dolosse on slopes steeper than 1:2 should be sustained by site-specific model tests.
[i]Applicable to slopes ranging from 1:1.5 to 1:5.
Source: From U.S Army Corps of Engineers [1984].

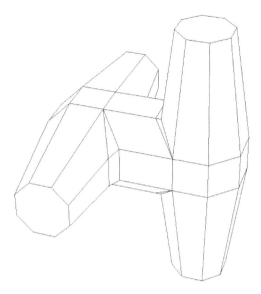

Fig. 5.9 An isometric view of Kolos.

5.6.2 *Dolos vs. Kolos*

The common failure modes for Dolos are found to be either at shank due to torsion dominated or at fluke by bending induced shear. In order to reduce the mid-shank failure due to flexure, the elongated shank length of 2862 mm in the Dolos is shrunken to a length of 2249 mm in Kolos. The volume of Kolos is kept equal to that of Dolos by making small modifications in other dimensions. Figure 5.10 clearly distinguishes between Dolos and Kolos. A profile of the Kolos and Dolos armour units with their characteristic dimensions is shown in Fig. 5.11.

5.6.3 *Finite element model*

Due to the complex geometry of the Kolos, the finite element method (FEM) of analysis is employed in this study using ANSYS multiphysics.

An isometric and plan view of the FEM model of Kolos and Dolos also are presented in Figs. 5.12(a), 5.12(b) and Figs. 5.12(c), 5.12(d), respectively. The model is comprised entirely of three-dimensional, 10 node tetrahedral element (SOLID 187) having three degrees of freedom at each node. It has quadratic displacement behaviour and is well suited to modelling irregular meshes. The basic approach used for the FEA is the method enabled by free meshing.

176 Coastal Engineering: Theory and Practice

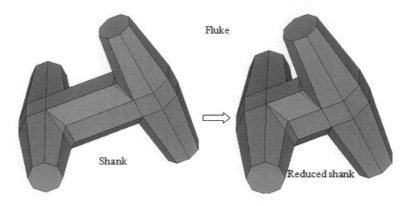

Fig. 5.10 Modification of Dolos into Kolos.

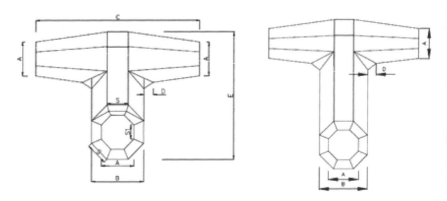

Particulars	Dimensions	
	Kolos	Dolos
A (mm)	600	572
B (mm)	960	916
C (mm)	2999	2862
D (mm)	171	163
E (mm)	2249	2862
S (mm)	399	379
S1 (mm)	249	237

Fig. 5.11 Definition sketch of Kolos and Dolos.

5. Rubble Mound Structures

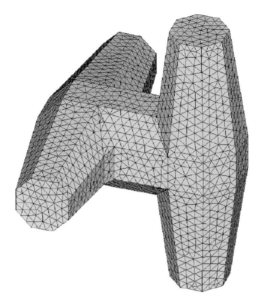

Fig. 5.12(a) An isometric view of the FEM discretization of KOLOS.

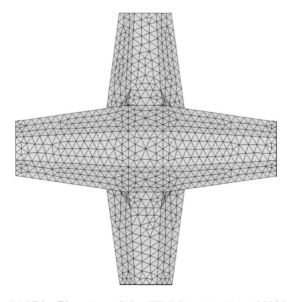

Fig. 5.12(b) Plan view of the FEM discretization of KOLOS.

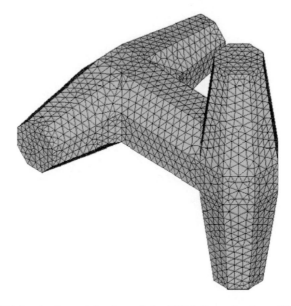

Fig. 5.12(c)   An isometric view of the FEM discretization of DOLOS.

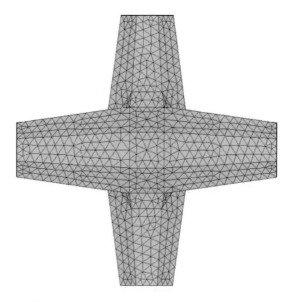

Fig. 5.12(d)   Plan view of the FEM discretization of DOLOS.

5.6.4 *Results of analysis*

Maximum tensile stresses from the FEM analyses are summarized in Table 5.2. Various load cases were analysed for Dolos and Kolos of equivalent weights. Figure 5.13 gives the comparison of maximum tensile stresses from FEM analyses.

Table 5.2. FEM static stress comparison.

Sl. No.	Boundary Condition	Loading Condition	Maximum stress (MPA) developed in the fillet region	
			Kolos	Dolos
1	The horizontal and vertical fluke bottoms are pinned	Flexure loading of two 4.5 ton both tips of horizontal fluke	0.832	0.909
		Torsional load of four 4.5 ton applied on all fluke tips	5.285	5.109
2	One of the fluke bottom is pinned horizontally	Flexure loading of two 9 ton tips of unsupported fluke	(−) stress has developed due to compression	
			−5.7	−6.5
		Flexure loading of 9 ton mid of unsupported fluke	3.922	6.582
		Flexure loading of 4.5 ton mid of unsupported fluke	*1.961*	*3.291*
		Torsional loading of two 4.5 ton unsupported fluke tips	*2.842*	*3.186*
3	One of the face of shank is pinned	Flexural loading of 9 ton one tip of horizontal fluke	2.605	6.971
		Torsional loading of 4.5 ton all fluke tips	8.624	6.311
4	Horizontal fluke is supported along its outer edge	(4a) Flexural loading of 9 ton vertical fluke tip	3.418	3.604
		(4b) Torsional loading of 4.5 ton vertical fluke	3.675	3.947
		Combination of load cases 4a and 4b	6.278	6.701

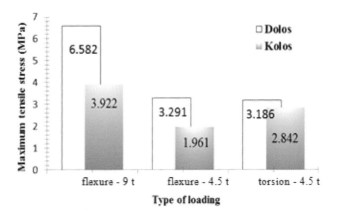

Fig. 5.13 Comparison of maximum tensile stresses from FEM analyses.

From the analyses conducted to understand the structural behaviour of kolos, it has been found that there is 50% reduction in the tensile stress due to bending in kolos unit compare with that of dolos. And for some load cases, kolos failed near the fillet regions because of the development of high stress concentration. Hence it is recommended to provide longitudinal and torsional reinforcement in order to take care of tension in those regions so that the structural stability of Kolos is enhanced. Concrete armour unit, KOLOS is economical and has better structural behaviour than dolos.

5.7 Stability of CAUs

The K_D value for any of the CAUs need to be assessed before it can be adopted in the field, the details of which are discussed herein considering KOLOS as the CAU. The stability of its armour layer has got the greatest significance. This is due to the fact that the failure of the armour layer would subsequently expose the relatively light weighted underlayer and core to the wave action, thus questioning the stability of the entire structure. The damage has hence been related to the armour and defined as the displacement of an armour unit and as per CEM [2006], the term "displacement" is either the units sliding over the slope or being dislodged from the layer. The stability parameters of KOLOS units placed in two layers were figured based on conducting two-dimensional physical model tests with an adopted scale, under the action of at least 1000 regular as well as random waves. The examination of stability coefficients involves the test criteria by changing the slope for both head and trunk section of breakwater armoured with KOLOS. The K_D value was evaluated using the well-known Hudson's formula both for the trunk and head sections.

The hydraulic stability of breakwater armour units is determined by wave height, unit size, specific density (of armour unit and sea water) and by the shape of the unit. The Stability Co-efficient (K_D) can be derived using the Hudson's formula given in Eq. (5.1). The Hudson's formula can be written in terms of stability number (N_s) as:

$$N_s = \frac{H}{\Delta D_n} = (K_D \cot \alpha)^{1/3} \qquad (5.2)$$

Δ = relative density = $[(\rho_c/\rho_w) - 1]$ and D_n = nominal diameter of armour unit ($V^{1/3}$).

Prior to obtaining the value of K_D for KOLOS, the damage level was fixed in line with the proposed guidelines on the classification of damage levels suggested by d'Angremond and van Roode [2004]. For an armour slope of 1 in 1.5, no repair is needed for an initial damage (no repair of armour) of up to 0.5% for DOLOS, irrespective of the weight of the individual unit of the armour layer and as KOLOS is an improved version of DOLOS, the same damage level has been incorporated in this stability study. The initial part of the study focuses on the establishment of the Hudson's stability coefficient of KOLOS, where the test sections had been subjected to the maximum wave height it can withstand without being damaged extensively. The maximum wave heights under the respective wave periods are illustrated in Table 5.3.

The Hudson's stability number has been established for KOLOS in order to have comparisons with other armour units and as from the table above, KOLOS can have a K_D value of up to 32, which is quite high, with a maximum damage percentage of 0.5%. Using such a high value for the design can be justified because random placement and double-layer configuration adds to the stability of the armour and even at a damage percentage of 0.5%, further infliction of damage cease to exist. The design K_D values of KOLOS for different armour slopes and for specific damage conditions are summarized in Table 5.4. It has to be noted down that these values corresponds to non-breaking wave conditions. The details of the hydrodynamic characteristics of KOLOS are discussed by Chandramohan et al. [2012].

Table 5.3. Maximum wave conditions during the tests (regular waves).

Armour slope	Wave period (s)	Wave height (m)	Damage % Damage	Damage Type	Stability coefficient (K_D)
1 in 1.5	1.2	0.21	0%	none	11*
	1.4	0.25	0%	none	19*
	1.6	0.3	0.3%	Roll down	32
	1.8	0.31	0.8%	Roll down	35
	2.0	0.32	0.3%	Roll down	39
1 in 2.0	1.2	0.22	0.7%	rocking	10*
	1.4	0.26	0.7%	rocking	16*
	1.6	0.31	0.24%	Roll down	27
	1.8	0.28	0.24%	Roll down	20
	2.0	0.31	0.24%	Roll down	27
1 in 2.5	1.2	0.22	0.37%	rocking	8*
	1.4	0.24	0.18%	rocking	10*
	1.6	0.32	0.18%	Roll down	23
	1.8	0.3	0.18%	Roll down	19*
	2.0	0.32	0.55%	Roll down	23

*Limited wave generation capacity, no instability.

Table 5.4. Recommended values of K_D for KOLOS (non-breaking waves).

	Hudson's stability coefficient for trunk (K_D)	
Trunk Slope	Zero Damage	Upto 0.5% Damage
1 in 1.5	23	32
1 in 2.0	19.5	27
1 in 2.5	14	23

5.7.1 Damage assessment

General

The total percentage of damage of any breakwater cross-section can be assessed either by measuring the damaged portion or by counting the number of displaced units. Both the methods are explained below.

Area method

The damage to the armour layer can be given as a percentage of displaced blocks related to a certain area (the whole or a part of the layer). Damage can be defined as the relative damage, N_o, which is the actual number of units displaced or rocking related to a width along the longitudinal axis of the structure of one nominal diameter D_n. An extension of the subscript in N_o can give the distinction between displaced out of the layer, units rocking within the layer (only once or more times). The actual number is related to a width of one D_n.

N_{od} = units displaced out of the armour layer (hydraulic damage)
N_{or} = rocking units and
$N_{omov} = N_{od} + N_{or}$.

5.7.2 Number of units method

This method damage assessment follows the procedure of Owen and Allsop [1983]. The damage to the groin section is in terms of the number of armour units, which have been totally dislodged from the armour layer. This can be expressed as a percentage of the total number of units on the armour layer. The movement of armour units can assess the damage and four categories of armour unit movements are given below:

- P — Unit seen to be rocking, but not permanently displaced.
- Q — Unit displaced by up to 0.5D
- R — Unit displaced between 0.5 and 1.0D
- S — Unit displaced by more than 1.0D

where, D is the equivalent diameter of an individual armour unit. The order of damage for armour layer has been assessed using the formula,

$$\% \text{ of damage} = \frac{\text{Number of armour units dislodged}}{\text{Total number of units in armour layer}} \times 100$$

The total is P+Q+R+S which it is less than 5%, the section is accepted to be safe. For the breakwater sections, the assessment of damage by the number of units method was adopted, in which case, the units are continuously monitored during the propagation of waves over the CAU. The study on the stability of breakwater sections and their analysis was based on visual observations of its physical behaviour. As stated in this section, the number of units' method is adopted for assessing damage of the rubble mound sections.

A view of the breakwater after establishing its K_D in the Laboratory in Department of Ocean Engineering in Krishnapatnam port situated along the east coast of India (14°15'N, 80°08'E) is projected in Fig. 5.14.

Fig. 5.14   A view of breakwater in Krishnapatnam port with Kolos as CAU.

References

John Ashlin, S., Sannasiraj, S. A. and Sundar, V. (2016). Hydrodynamic performance of an oscillating water column device exposed to oblique waves, *Proc. 12th International Conference on Hydrodynamics*, TU Delft, Netherlands.

Chandramohan, P. V., Sundar, V., Sannasiraj, S. A. and Arunjith, A. (2012). Development of 'KOLOS' armour block and its hydrodynamic performance, *Proc. 8th Int. Conf. on Coastal and Port Engineering in Developing Countries*, pp. 1344–1354.

Coastal Engineering Manual (2006). Part VI, U.S. Army Engineer Waterways Experiment Station, U.S. Government Printing Office, Washington, DC.

d'Angremond, K. and van Roode, F. C. (2004). *Breakwaters and Closure Dams*, SPON Press, London.

Markle, F. G. and Davidson, D. D. (1979). Stability coefficients for placed-stone jetties, *Engineering Technical Letter 1110-2-242*.

Owen, M. W. and Allsop, N. W. H. (1983). Hydraulic modelling of rubble mound breakwaters, *Proc. Cong. Breakwaters, Design and Construction*, ICE, London.

Sasajima, H., Koizuka, T. and Sasayama, H. (1994). Field demonstration test on a semi-circular breakwater, Hydroport '94, Yokosuka, Japan 1, pp. 593–615.

Sundar, V., Moan, T. and Hals, J. (2010). Conceptual designs on integration of oscillating water column devices in breakwaters, *Proc. OMAE 2010*.

Zwamborn, J. A. and Van Niekerk, M. (1982). Additional model tests. Doly packing density and effect on relative block density, National Research Institute of Oceanography, Coastal Engineering and Hydraulic Division, Stellenbosh, South Africa, July.

Chapter 6

Wave Run-up and Overtopping

6.1 Introduction

Along a coast, the water levels tend to change in time based up on the tidal fluctuations and wave climate which comprises of two components. One is dynamic about the mean water level due to the wave action along the shore and another is the static with gradually varying of water level associated with wave set up, astronomical tides and storm surge (abnormal rise of water level). Wave deformation occurs when it propagates from the offshore to shallow water, and when it interacts with coastal structures like breakwaters, sea walls, revetments, groins and dykes it run-ups and sometimes flows over the structure or the coast, which is termed as overtopping. The physical processes of wave run-up and overtopping are the two major complex phenomena that influence the design of seawalls exposed to wave attack. In the design of breakwater, the crest elevation from the still water level is established based on the maximum wave run-up height. The wave run-up and overtopping are influenced by factors like geometrical parameters such as slope, free board and crest width, structural parameters that includes porosity, permeability, stone shape, size and layer thickness, and hydraulic parameters such as mean sea level, celerity, wave height distribution, spectral shape, wave direction, wave grouping, wave period, wave breaking and currents.

The wave run-up and overtopping significantly affects the functional efficiency which involves protection of shipment, and, to a certain extent, the structural stability of coastal structures as well as coastal flooding. Though, the overtopping sometimes may not be due to a structure, the splash of water by wind carried spray may cause inconvenience to the residents close to the area on the landward side along the coast. On the lee side of low crest breakwaters, the said phenomena lead to damages of the reclaimed areas. In such cases, breakwaters with higher crest elevation should be adopted to minimize the overtopping and to provide good shelter, on the contrary, this involves higher investment as well as it affects the aesthetics. As the ocean waves are stochastic in nature, a mathematical expression for these two factors are not possible. Hence, the wave run-up and overtopping for coastal structures are established based on the experimental investigations.

6.2 Wave Run-up

6.2.1 General

The wave run-up is the maximum vertical extent of wave up rush on a beach or structure above the still water level (SWL) [Sorensen, 1997]. Figure 6.1 describes the limit of wave run-up on a beach. It governs the level over which, the waves influences. It is therefore a significant parameter to define the flood inundation ranges due to storms. Longshore and cross-shore sediment transport are related to wave run-up as it is a function of the hydrodynamics on the beach.

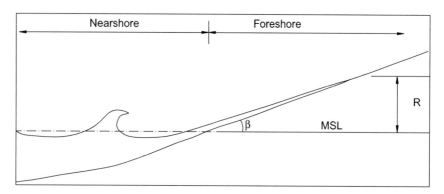

Fig. 6.1 Wave run-up.

Over the past few decades, wave run-up has been extensively studied. The research studies also includes Kobayashi [1999], the Coastal Engineering Manual [2002], and recently, the EurOtop Manual [2007]. Wave run-up can vary according to the characteristics of the structure, like permeable core of which rubble mound breakwaters, impermeable core of revetments and levees, smooth grass and rough impermeable concrete armoured structures with uniform or complex slopes. On the other hand, beaches are categorized by an average slope ranging between 1:100 to 1:10. The beach morphology undergoes variations in time with beach steepening during a storm activity when the sediments and the near shore bar move far away from shore. Hence, the run-up estimation is uncertain with such variations in the beach slope. Steeper beaches are wave-reflecting, while, shallower-sloping beaches are wave-dissipating.

According to Stockdon et al. [2006], wave run-up involves two parts, viz., wave setup which is a mean water surface elevation averaged over time and swash. Hence, wave run-up equation incorporating swash, S, which is the change of water-land boundary about the MSL is shown below.

$$R = \eta_{\max} + \frac{S}{2} \tag{6.1}$$

where, $\eta_{\max}$ is the maximum wave setup, which is the super elevation of the mean water level at the shoreline. One of the early formulae for run-up of breaking monochromatic waves on a plane slope was proposed by Hunt [1959] which is,

$$\frac{R}{H} = \frac{\tan \alpha}{\sqrt{H/L_0}} \tag{6.2}$$

where α is the slope angle (i.e., $\tan \alpha = m$). The right-hand side of the above equation is the Iribarren number [Iribarren and Nogales, 1949] or surf similarity parameter (ξ).

Hunt [1959] summarized the useful data based on analytical and experimental study concerning run-up and overtopping in connection to the walls with an inclined seaward faces of simple or composite form, either with or without berms, encountered by non-breaking waves. The most important conclusion with respect to the slope of the sea wall face, which is to ensure breaking of wave is given by,

$$\tan \alpha = \frac{8}{T} \left(\frac{H_i}{2g} \right)^{\frac{1}{2}} \tag{6.3}$$

Such a slope will result in the reflected wave being approximately 50% of the incident wave height. Therefore, the minimum slope, or nevertheless the level of apron up to the breaking point of wave, should be resolute in relation to the lowest wave frequency of critical height. Another conclusion was that the run-up, R, of a breaking wave, measured vertically above the MSL at any given time, may be related to the incident wave height H_i such that,

$$\frac{R}{H_i} = \frac{KT \tan \alpha}{8} \left(\frac{H_i}{2g} \right)^{-\frac{1}{2}} \tag{6.4}$$

where, K is a constant of 2.3 for a smooth place surface and for a surging wave, this ratio will be no greater than 3 and has been shown theoretically that in the absence of friction.

$$\frac{R}{H_i} = \left(\frac{\pi}{2\alpha}\right)^{\frac{1}{2}} \quad \text{for } \frac{\pi}{4} < \alpha < \frac{\pi}{2} \tag{6.5}$$

Table 6.1. Roughness coefficients for different surface textures (various sources) as reported by Reeve et al. [2012].

Type of slope protection	Roughness coefficient, r	Roughness factor from EurOtop, γ_f
Smooth concrete or asphalt	1.0	1.0
Smooth concrete blocks with little or no drainage	1.0	1.0
Stone blocks pitched or mortared	0.95	
1/4 stone set 10 cm higher	0.9	
Stepped	0.9	
Turf	0.85–0.9	1.0
Basalt		0.9
Rough concrete	0.85	
Small blocks over 1/25 surface		0.85
Small blocks over 1/9 surface		0.8
One layer of rock armour or an impermeable base	0.8	0.60
Open stone asphalt	0.8	
Stones set in cement, rag stone etc.	0.75–0.8	
Ribs (spacing /width = 7, height/width = 5–8)		0.75
Fully grouted stone	0.6–0.8	
Partially grouted stone	0.6–0.7	
Rounded stones	0.6–0.7	
One layer of rock armour on a permeable base	0.55–0.60	0.45
Two layers of rock armour	0.5–0.6	0.55 (impermeable core) 0.40 (permeable core)
Hollow cube armour units, one layer	0.5	
Antifiers		0.47
Dolos and Accropode^{TM} armour units	0.4	0.46
Xbloc^R		0.45
CORE-LOC^R		0.44
Stabit armour units	0.35–0.4	
Tetrapods, two layers	0.3	0.38

It is a common practice to express the effect of surface texture of the slope's surface to a coefficient of roughness defined as the ratio of the rough to smooth run-up. The roughness coefficient for the different types of textures of the slopes with or without the armour layers are provided in Table 6.1. For composite slopes Saville [1958] suggested that a rational approximation of run-up could be related to a corresponding slope that traverses the actual slope at the position of the breaker point and the maximum run-up. This requires an iterative solution to resolve and will under-estimate the run-up measured for concave slopes. The introduction of a berm into a slope can provide a very effective means of reducing run-up provided the width of the berm represents a significant part, say 20% of the wavelength. This is typically of the order of 10 m for shallow water coastal defence structures. A berm is also generally most effective when positioned at or above the MSL. In an environment exposed to a high tidal range, the definitive level will often be taken as mean high water springs or that determined from a joint wave and water level probability analysis.

Battjes [1974] extended the formula of Hunt [1959] for regular to irregular waves using the time-domain wave parameters. The formula for the relative wave run-up is as follows:

$$\frac{R_{2\%}}{H_s} = C_m \xi_m \tag{6.6}$$

$$\xi_m = \frac{\tan \alpha}{\sqrt{S_m}} \tag{6.7}$$

$$S_m = \frac{H_s}{L_m} \tag{6.8a}$$

$$L_m = \frac{gT_m^2}{2\pi} \tag{6.8b}$$

where, $C_m = 1.49$–1.87, $R_{2\%} =$ Run-up exceeded by 2% of the number of incident waves.

$\alpha =$ slope of the structure and $T_m =$ mean wave period.

Another set of run-up data for smooth slopes is presented by Waal and van der Meer [1992] for, $\xi_m > 2.0$,

$$\frac{R_{u2\%}}{(H_s)_{toe}} = 3.0 \tag{6.9}$$

Ahrens [1988] developed an empirical formula to calculate the maximum run-up due to irregular waves over riprap embankments. The method was based on energy-based zero moment wave height. It was found that the maximum run-up was not too sensitive to the armour layer thickness.

$$\frac{R_{\max}}{H_s} = \frac{a\xi_m}{1.0 + b\xi_m} \qquad (6.10)$$

where, "ξ_m" is the surf similarity parameter for random waves, a and b are the dimensionless run-up coefficients determined by regression analysis.

The equation proposed by EurOtop [2007] is given as

$$\frac{R_{2\%}}{H_s} = 1.65\gamma_b\gamma_f\xi_m \qquad (6.11a)$$

$$\frac{R_{2\%}}{H_s} \leq \gamma_b\gamma_{fsurging}\gamma_\beta(4 - 1.5/\sqrt{\xi_m}) \qquad (6.11b)$$

$\xi_m = 1.8$ the roughness factor $\gamma_{fsurging}$ increases linearly up to 1 for $\xi_m = 10$, which can be described by,

$$\gamma_{fsurging} = \gamma_f + (\xi_m - 1.8)(1 - \gamma_f)/8.2 \qquad (6.12)$$

and

$$\gamma_{fsurging} = 1.0 \quad \text{for } \xi_m > 10 \qquad (6.13)$$

where,

γ_b = correction factor for a berm
γ_f = correction factor for the permeability and roughness of or on the slope
γ_β = correction factor oblique wave attack
ξ_m = breaker parameter that considers the combination of structure slope and wave steepness. For complete details refer to EurOtop manual [2007].

An important objective of the design of the most types of coastal structure founded on a soft or erodible seabed is to maximize the destruction of wave energy by initiating the wave to break on the wall. Furthermore, by an appropriate choice of outer layer or wall profile and to an extent the degree of surface roughness with an intention to endorse maximum turbulence of the swash over the surface of the structure is also in practice.

Van der Meer and Stam [1992] estimated the wave run-up over smooth and rock slopes due to irregular waves through experimental investigations.

It was reported that the run-up reaches a maximum for impermeable structures, whereas, the permeability of the structure has an influence on the run-up only for lower values of ξ_m.

$$\xi_m = \tan \alpha \left/ \sqrt{\frac{2\pi H_s}{T_m^2}} \right. \qquad (6.14)$$

For larger values of (i.e., "ξ_m" > 6) smooth and rock slopes were found to yield the same levels of run-up. The formulae for the 2% run-up level for the mean wave period for a double layered rubble mound slope were proposed as follows,

$$R_{u2\%}/H_s = 0.96\xi_m \quad \text{for } \xi_m < 1.5 \qquad (6.15)$$

$$R_{u2\%}/H_s = 1.17\xi_m^{0.46} \quad \text{for } \xi_m > 1.5 \qquad (6.16)$$

6.2.2 Recent run-up equation

For plane progressive waves, based on the maximum depth-integrated wave momentum flux Hughes [2003, 2004] presented a new non-dimensional parameter. The parameter mentioned as the wave momentum flux parameter (P_{MF}), was well-defined as,

$$P_{MF} = \left(\frac{M_F}{\rho g d^2}\right)_{\max} \qquad (6.17)$$

where

M_F = depth-integrated wave momentum flux

As $(M_F)_{\max}$ consists of force per unit length of wave crest, it was claimed that maximum depth integrated wave momentum flux would deliver a good representation of wave processes at coastal structures.

For establishing an empirical equation for estimating the wave momentum flux parameter for finite amplitude, Hughes [2003, 2004] considered, non-linear waves based on a numerical solution technique (Fourier approximation) and the resulting empirical equation, was given as,

$$\left(\frac{M_F}{\rho g d^2}\right)_{\max} = A_0 \left(\frac{d}{gT^2}\right)^{-A_1} \qquad (6.18)$$

$$A_0 = 0.639 \left(\frac{H}{d}\right)^{2.026} \qquad (6.19)$$

$$A_1 = 0.180 \left(\frac{H}{d}\right)^{-0.391} \qquad (6.20)$$

Detailed information and example calculation of wave momentum flux M_F, are specified in CHETN-III-67. A simple generic equation for wave run-up in terms of wave momentum flux was established by Hughes [2003]. It was claimed that the weight of fluid within the area ABC of Fig. 6.2. is proportional to the maximum depth-integrated wave momentum flux of the wave just prior to it reaching the toe of the structure slope. The resultant equation was given by the form,

$$\frac{R}{d} = C \cdot F(\alpha) \left(\frac{M_F}{\rho g d^2} \right)^{0.5} \qquad (6.21)$$

where $C \cdot F(\alpha)$ is an unknown function of the slope of the structure. For convenience, the subscript "max" has been dropped from the wave momentum flux parameter.

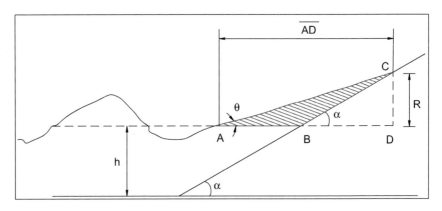

Fig. 6.2 Maximum wave run-up on plane impermeable slope.

In the run-up equation (6.21), the relative run-up (R/d) is directly proportional to the square root of the wave momentum flux parameter. (Note that representing the run-up sea surface slope as a straight line is an approximation and may not be fully appropriate for non-breaking waves on steep slopes, wherein, the water surface has pronounced curvature.)

Based on the laboratory and field studies for different types of waves over different slopes, the wave run-up established by researchers are tabulated in Table 6.2.

Table 6.2. Wave run-up methods.

Method	Slopes	Waves	Setting	Equations	Comments
Hunt [1959], Battjes [1974]	Smooth, impermeable, continuous	Regular	Laboratory	$\dfrac{R}{H_0} = \xi$	$0.1 < \xi < 0.3$
CERC [1984]	Smooth, impermeable, continuous	Regular	Laboratory	Chart solution, refer CEM [1984]	$\dfrac{d_s}{H_0} \approx 2.0$
Mase [1989]	Smooth, impermeable, continuous, gentle: $\theta = 1.9°$–$11.3°$	Irregular	Laboratory	$\dfrac{R}{H_0} = a\xi^b$, $R_{2\%}: a = 1.86, b = 0.71$ $\bar{R}: a = 0.88, b = 0.69$	$0.007 \leq \dfrac{H_0}{L_0} \leq 0.07$
Nielsen and Hanslow [1991]	Beach sand of $\theta = 1.5°$–$10.8°$ Mean grain size $= 0.18$–$0.8\,\text{mm}$	Irregular	Field	$R = cL_{swm}, R_{2\%}: c = 1.98$ $L_{swm} = 0.6(H_{orms}L_0)^{0.5}\tan\beta_F$ $L_{swm} = 0.05(H_{orms}L_0)^{0.5}$	for $\tan\beta_F \geq 0.10$ for $\tan\beta_F < 0.10$
Ahrens and Seelig [1996]	Sand and gravel beaches	Irregular	Laboratory and Field	$\dfrac{R_{2\%}}{H_{so}} = \dfrac{4.1}{N_0}\sqrt{\dfrac{d_{sw}}{d_{sr}}}$ $N_0 = \dfrac{H_{so}}{w_{sr}T}$ $w_{sr} = 14.5(d_{sr})^{1.1}$	d in mm, w_{sr} in cm/s w_{sr} for fresh water and $0.15 \leq d_{sr} \leq 0.85\,\text{mm}$

6.3 Wave Overtopping

6.3.1 *General*

The overtopping of water over a coastal structure depends on its crest elevation. As overtopping dictates the crest elevation, it also influences the cost effect of a structure. Curiously, it is not something that has attracted a significant amount of research until recently. However, the modern practice of design involves the assessment of overtopping as a design criteria instead of wave run-up, which does not quantify the discharge over a structure, whereas, the run-up approach as design criteria was factually used due to the lack of adequate design data.

Conventional design practices require a structure to protect against critical wave climate that would damage the structure as well as spoils the beach or land and also affects the future development that follows. The latter will often make it necessary to consider good standards of design practice for safety reasons and also to avoid the damage to a property than would be required for protection of the structure along. Such features could be other structures, such as buildings, roadways or working areas as in the case of port facilities. It is necessary to consider the safety of people or vehicles behind the structure. However, designing to meet these criteria can lead to the development of tremendously huge erections. Consequently, it is only during the exceptional circumstances of critical conditions these criteria comes into consideration for example when a highway present immediately behind a sea wall, densely populated coastal community. Although, the allowable discharge over a structure is small, it is worth analyzing that it could result in localized flooding, and duration with level of depth of flooding could be the governing factor in determining the appropriate level of overtopping discharge. Tolerable discharges over a structure can be established through the knowledge of drainage capacity, and also fixing the spreading area of the flood thereby limiting the depth which would in turn limit the allowable volume per unit meter of defense. Further, the actual discharge can be calculated as a whole volume instead of a mean rate and calculating the incremental volume due the action of tidal variations. These are the primary reasons for limiting overtopping discharge and it is therefore important to establish the design criteria that relate to all facets of the structure and its intended performance.

Overtopping discharges are usually calculated and quoted as mean discharges (litres/sec/m run) and can appear to be relatively small values. However, the actual discharge occurs as a random series of large single — impact events (i.e. every wave crest) with a frequency equal to the wave period. It should also be realized that overtopping calculation methods have limitations to their accuracy and the physical model data from which the methods are derived generally exhibit considerable scatter. It is generally accepted that even the most reliable methods cannot provide absolute discharges, and they can only be assumed to produce overtopping rates that are accurate to within one order of magnitude. Likewise, the tolerable discharges defined in various publications should not be taken as absolute values. They represent an order of magnitude for which damage or unsafe conditions may exist.

6.3.2 Calculation of overtopping rates

Some of the initial data on calculating the overtopping rates was undertaken in the 1950s, the results of which are presented in the Shore Protection Manual [SPM, 1984]. Later this information was outmoded by the work carried out by a number of researchers, most remarkably Owen [1980] who established the formulation outline that continues to be used today. The most recent definitive and comprehensive work, which reports overtopping for various forms of structures, and has been taken over and published by Besley [1999].

The mean discharge due to overtopping over a plain rough-armoured slope may be calculated from the following equation

$$R_* = \frac{R_c}{T_m(gH_s)^{0.5}} \qquad (6.22)$$

where, R_c is the freeboard from the still water level to the top of the crest elevation and Eq. (6.22) is valid only between the limits of $0.05 < R_* < 0.30$. The second parameter is shown as,

$$Q_* = A \exp\left(-\frac{BR_*}{r}\right) \qquad (6.23)$$

whereas, B is the empirical coefficient which depends on the slope of the structure slope (see Table 6.3) and r is the roughness coefficient as given in Table 6.1. This equation is valid in the range $0.05 < R_* < 0.30$. The average overtopping discharge rate per meter length of the structure in m^3/s/m is

$$Q_m = Q_* T_m g H_s \qquad (6.24)$$

Table 6.3. Empirical coefficients – simply sloping sea walls [Besley, 1999].

Sea wall slope	A	B
1:1	7.94E – 3	20.1
1:1.5	8.84E – 3	19.9
1:2	9.39E – 3	21.6
1:2.5	1.03E – 2	24.5
1:3	1.09E – 2	28.7
1:3.5	1.12E – 2	34.1
1:4	1.16E – 2	41.0
1:4.5	1.20E – 2	47.7
1:5	1.31E – 2	55.6

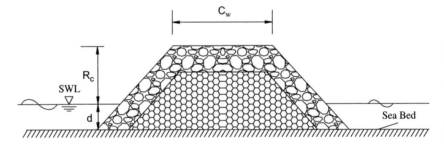

Fig. 6.3 Definition sketch for wave overtopping rough plane slope.

If the structure has a permeable crest berm, a reduction factor C_r may be applied as

$$C_r = 3.06 \exp\left(-\frac{1.5 C_w}{H_s}\right) \qquad (6.25)$$

Where C_w is the crest width as indicated in Fig. 6.3. If C_w/H_s is less than 0.75 it can be assumed that the reduction is zero.

For waves approaching at an angle to the slope, a further reduction factor may be applied based on investigations be Banyard and Herbert [1995]. As indicated in the previous section, the introduction of a berm into a slope can be a very effective means of reducing the crest level to a lower elevation that would be required for a simple plane slope for the same overtopping discharge. This can often be important for aesthetic reasons or facilities, where, a line of sight over the structure is a key feature. Besley [1999] proposed that for a slope with a berm that is below the still-water level, Eq. (6.25) can be used together with modified empirical coefficients given in Table 6.3 together with the slope of the upper section of the structure as indicated in Fig. 6.4. For berms above the MSL, it is suggested that an equivalent slope based on the plane that joins the intersection of the lower slope with the SWL and the top of the seaward slope of the upper section as shown in Fig. 6.4 should be used. Then Eq. (6.22) to Eq. (6.25) may be used in conjunction with the empirical coefficients given in Table 6.3 that most closely fit the equivalent slope.

Another method of reducing the height of a structure is to include a wave return wall. Coastal defense structures can therefore sometimes incorporate a return wall for waves either on the top of the slope directly or at some distance retired from the crest of the seaward slope. A comprehensive study was carried out by Owen and Steele [1991] who evaluated the performance of wave return walls, in terms of a discharge factor D_f which was defined as the ratio of the discharge overtopping the re-curve wall to the equivalent discharge in the absence of the wall. Referring to Fig. 6.5, Eqs. (6.22) through (6.25) apply in order to determine Q_m which is the discharge per

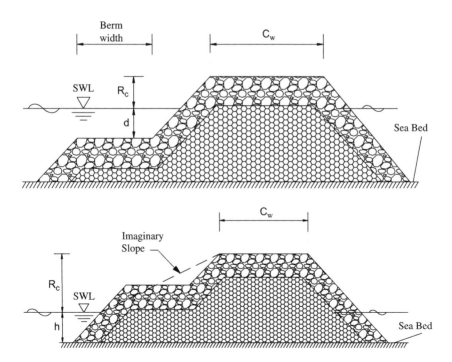

Fig. 6.4   Definition sketch for bermed sea wall.

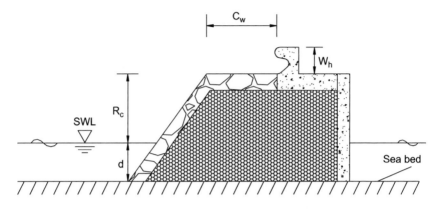

Fig. 6.5 Definition sketch for wave return wall.

metre run (m³/s/m) at the base of the return wall and is the same as that at the crest of the slope for a smooth impermeable crest.

A and B are the same empirical coefficients given in Table 6.4. A dimensionless wall height is defined as,

$$W_* = \frac{W_h}{R_c} \qquad (6.26)$$

Where W_h is the height of the wave return wall and R_c is the freeboard to the top of the wall as previously defined. From here on the procedure for impermeable and impermeable structures differs. For an impermeable structure, given the dimensions wall height, the seaward slope of the sea wall and the set-back distance of the wave return wall, Table 6.5 provides values of an adjustment factor A_f which in terms used to define an "adjusted slope freeboard" given by:

$$X_* = A_f R_c \qquad (6.27)$$

Figure 6.6 is then used to determine a discharge factor (D_f) for the given conditions so that the mean discharge over the wall is:

$$Q_W = Q_m D_f \qquad (6.28)$$

The overtopping discharge can both increase significantly and decrease depending on the wave approach angle. The ratio of the discharge under angled wave attack to perpendicular attack, O_r in the case is given as:

$$O_r = -1.18 In(D_f) - 0.40 \qquad (6.29)$$

Table 6.4. Empirical coefficients — bermed sea walls — berm at or below SWL (after Besley [1999]).

Sea wall slope	Berm elevation (m)	Berm width (m)	A	B
1:1	−4.0	10	6.40E − 3	19.50
1:2			9.11E − 3	21.50
1:4			1.45E − 2	41.10
1:1	−2.0	5	3.40E − 3	16.52
1:2			9.80E − 3	23.98
1:4			1.59E − 2	46.63
1:1	−2.0	10	1.63E − 3	14.85
1:2			2.14E − 3	18.03
1:4			3.93E − 3	41.92
1:1	−2.0	20	8.80E − 4	14.76
1:2			2.00E − 4	24.81
1:4			8.50E − 3	50.40
1:1	−2.0	40	3.80E − 4	22.65
1:2			5.00E − 4	25.93
1:4			4.70E − 3	51.23
1:1	−2.0	80	2.40E − 4	25.90
1:2			3.80E − 4	25.76
1:4			8.80E − 4	58.24
1:1	−1.0	5	1.55E − 2	32.68
1:2			1.90E − 2	37.27
1:4			5.00E − 2	70.32
1:1	−1.0	10	9.25E − 3	38.90
1:2			3.39E − 2	53.30
1:4			3.03E − 2	79.60
1:1	−1.0	20	7.50E − 3	45.61
1:2			3.40E − 3	49.97
1:4			3.90E − 3	61.57
1:1	−1.0	40	1.20E − 3	49.30
1:2			2.35E − 3	56.18
1:4			1.45E − 4	63.43
1:1	−1.0	80	4.10E − 5	51.41
1:2			6.60E − 5	66.54
1:4			5.40E − 5	71.59
1:1	0.0	10	8.25E − 3	40.94
1:2			1.78E − 2	52.80
1:4			1.13E − 2	68.66

Further analysis of wave approach angle has identified that the large increases only occur for very small discharge factors and decreases only occur for discharge factors greater than about 0.3. The recommended approach is to use the "worst case" combination of D_f and O_r. Where

O_r is the wave incidence reduction factor. This translates into using D_f only when $D_f \geq 0.3$, and using $(D_f O_r)$ when $D_f < 0.3$.

For roughened slopes or those incorporating a berm, Besley [1999] recommends the determination of a smooth slope that gives the same overtopping discharge at the top of the slope, for the same wave conditions. That "equivalent slope" is then used to obtain the adjustment factor from Table 6.5. However, this may well produce slopes that lie outside the range of available data. The alternative is to calculate the overtopping using the method of Van der Meer [1998] described later.

For wave walls on permeable slopes Besley [1999] re-analyzed the data from Bradbury and Allsop [1988]. The base discharge is calculated in the same way as that described for permeable crests in Eqs. (6.20) and (6.21). Given W_* as defined in Eq. (6.26), the discharge factor is obtained directly from Fig. 6.6, so that the mean overtopping discharge becomes,

$$Q_w = Q_m C_r D_f \tag{6.30}$$

For plain vertical walls Besley [1999] summaries the work of Allsop *et al.* [1995]. A parameter h_* is defined as,

$$h_* = \left(\frac{d'}{H_s}\right)\left(\frac{2\pi d'}{gT_m^2}\right) \tag{6.31}$$

Table 6.5. Adjustments factors — wave return walls on impermeable sea walls (after Besley [1999]).

Sea wall slope	Crest berm width (C_w)m	A_f
(a) $W_* = W_h/R_c \geq 0.6$		
1:2	0	1.00
1:2	4	1.07
1:2	8	1.10
1:4	0	1.27
1:4	4	1.22
1:4	8	1.33
(a) $W_* = W_h/R_c < 0.6$		
1:2	0	1.00
1:2	4	1.34
1:2	8	1.38
1:4	0	1.27
1:4	4	1.53
1:4	8	1.67

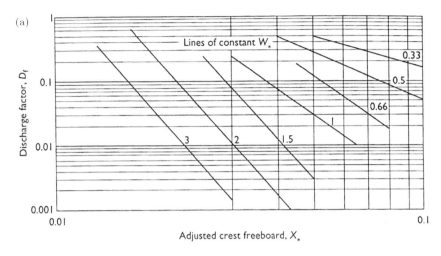

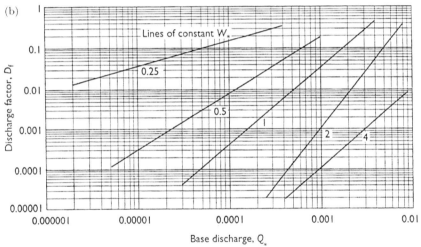

Fig. 6.6 Discharge factors with wave return walls for (a) impermeable slopes and (b) permeable slopes [Besley, 1999].

Where d' is the depth of water at the toe of structure for which reflecting waves dominate when $h_* > 0.3$ and impacting waves when $h_* < 0.3$. For the former, the mean overtopping discharge per metre run of wall is given as:

$$Q_m = 0.05 \exp\left(\frac{-2.78 R_c}{H_s}\right)(gH_s^3)^{0.5} \qquad (6.32)$$

Where R_c is valid in the range $0.03 < R_c/H_s < 3.2$. For angled wave attack, the reduction factor is:

$$O_r = 1 - 0.006\beta \quad \text{for } 0° < \beta < 45° \tag{6.33a}$$

$$O_r = 0.72 \quad \text{for } \beta > 45° \tag{6.33b}$$

For impacting waves:

$$Q_m = 0.000137 \left(\left(\left(\frac{R_c}{H_s}\right)h_*\right)^{-3.24}\right) h_*^2 \left(gd'^3\right)^{0.5} \tag{6.34}$$

Which is valid in the range $0.05 < (R_c/H_s)h_* < 1.00$. There is no equivalent expression for different angle wave attack. Besley [1999] also provides empirical expressions for composite vertical walls fronted by a mound that may be submerged or emergent.

6.3.3 Complex slopes

The work by Van der Meer [1998] is also considered to be of use for applications that lie outside the range covered by the foregoing methods, for example assessing overtopping of rather flatter and composite structures. In some circumstances, this method might be applied to beaches that have long been an unresolved problem. Discussing with the complex slope Fig. 6.7, the basic expression for the average overtopping rate is given as,

$$\frac{Q_m}{(gH_s^3)^{0.5}} = 0.06 \left(\tan \alpha\right)^{0.5} f_b \xi_s \exp\left(\frac{-4.7\left(\frac{R_c}{H_s}\right)}{(\xi_s f_b f_f f_o f_w)}\right) \tag{6.35}$$

With a maximum of

$$\frac{Q_m}{(gH_s^3)^{0.5}} = 0.2 \exp\left(\frac{-2.3 R_c}{(H_s f_b f_f)}\right)$$

A breaker parameter ξ_m is based on the geometry of the structure and a quasi-wave steepness parameter such that,

$$\xi_m = \tan \alpha \left(\frac{gT_m^2}{2\pi H_s}\right)^{0.5}, \quad \tan \alpha = \frac{3H_s}{(L_{slope} - B)} \tag{6.36}$$

And f_b, f_f, f_o and f_w are reduction factors for a berm, roughness friction, angle of wave attack and the presence of a vertical wall, respectively. The berm width is defined as the flat part of the profile that has a slope of less than 1:15. The effectiveness of a berm is dependent on its level in

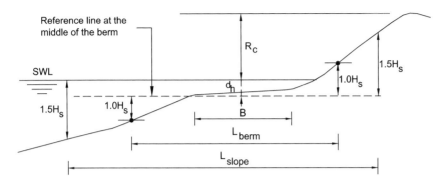

Fig. 6.7 Sketch of complex slope.

relation to the still-water level at the midpoint of the berm which defines the depth of water d_h. The berm reduction coefficient is given as,

$$f_b = 1 - \left(\frac{B}{L_{berm}}\right)\left(1 - 0.5\left(\frac{d_h}{H_s}\right)^2\right) \tag{6.37}$$

with $0.6 \leq f_b \leq 1.0$ and $-1.0 \leq d_h/H_s \leq 1.0$.

The effectiveness of the berm will be at its most when it is in still water level, hence an optimum width can be achieved when the reduction factor of the berm hits a value of 0.6. Therefore, the effective width of berm at SWL becomes,

$$B = 0.4 L_{berm} \tag{6.38}$$

The roughness reduction factor f_f is the same as those given in Table 6.1 although, Van der Meer [1998] does give some slightly different values in his publication for some types of roughness. For the angle of wave attack factor, it is suggested that for long-crested waves there is virtually no reduction within the range $0° < \beta < 30°$, but thereafter reduces fairly quickly to 0.6. However, for short-crested waves, which are usually more relevant to extreme conditions, the reduction factor for overtopping is given as,

$$f_o = 1 - 0.0033\beta \tag{6.39}$$

For the reduction factor for the presence of a small vertical wall or steep slope at the top of the wall an expression is provided in the original publication, but it is strictly limited to a restricted range of geometry and does not have general applicability.

6.3.4 Designing for overtopping

The generally accepted values for limiting values of mean overtopping rates are presented in CIRIA and CUR [1991] and McConnell [1998]. They are reproduced in diagrammatic form in Table 6.6. The following comments relate to the application of Table 6.6.

- The definition of "unprotected" is derived from a reference to a concrete revetment/pavement, the latter referring to compacted soil, grass or clay.
- The Dutch have different criteria that can be adopted which are more stringent [Van der Meer 1998]. However, it is not suggested these should be universally applied without more detailed analysis of acceptable risks at any location.
- A common misconception is that overtopping discharges reduce by an order of magnitude for every 10 m behind the crest or wave wall of a structure. There is a misinterpretation of a statement by Owen [1980]. He stated that the *limits* for *damage* (and therefore safety) could be increased by a factor of 10, at a distance 10 m behind the crest or wave wall. This is important with respect to flooding aspects as clearly the volume of water overtopping the crest or wave wall is a function of the wave conditions and the geometry of the structure.
- In terms of safety (and only this), Besley [1999] proposes the calculation of maximum or peak discharges. This is logical and it is recommended that this approach is adopted. This concludes that a structure becomes unsafe for pedestrians, if the overtopping rate surpasses $0.04\,\mathrm{m^3/m}$ added to that a structure becomes unsafe for vehicles driven at any speed when the overtopping incident surpasses $0.06\,\mathrm{m^3/m}$.
- Breakwaters are not usually designed for the above tolerable overtopping discharges unless they have facilities located on or directly behind them, or require frequent access, in which cases the above considerations apply. This is generally because the size of armour on the crest and rear slope is much larger than that considered as "protection" by the limits specified above.

Whilst work has been conducted to investigate tolerable overtopping discharges, little has been carried out to establish the size of protective cover layers to avoid overtopping damage. Guidance is given by Pilarczyk [1990] and reproduced in CIRIA and CUR [1991] although the original reference provides more background. This provides a method for calculating both the size of rock required and the width of protection, although the

Table 6.6. Acceptable overtopping limits [Owen, 1980].

Functional Safety			Structural safety		
				Damage even for paved promenade	200
100 —	Unsafe at any speed	Very dangerous	Structural damage	Damage if fully protected	
				Damage even for paved promenade	
10 —					50
		Grass like dangerous		Damage if back slope not protected	
					20
	Horizontal composite wall unsafe to park	Horizontal composite wall dangerous		Damage if crest not protected	
0.1					2
					0.6
	Vertical wall unsafe to park	Vertical wall dangerous		No damage	0.3
					0.03
0.01		Uncomfortable but not dangerous	Minor damage to fittings, etc.	No damage	
	Unsafe at high speed				0.004
0.001	Safe at all speeds	Wet, but not uncomfortable	No Damage		
	Vehicles	Pedestrians	Buildings	Embankment sea walls	Revetment sea walls

former requires calculation of run-up levels and the latter includes a factor related to the important of the structure without specific guidance on what this value should be. Consequently, the methods require some degree of interpretation. A crest width should be a minimum of three primary armour stone widths, i.e. $3D_{n50}$ is recommended. Although as a conventional thumb rule, Pilarczyk also recommended that the crest and lee slope may be protected over a width equal to the projected extent of run-up beyond the crest of the structure.

In order to avoid ambiguity, the general descriptions used in setting the tolerable discharges need to be broadly applied so that there distinctions made, for example, between turfed/compacted gravel and format protection such as armour stone or concrete blocks.

6.4 Summary

Guidance for wave run-up and overtopping provided by EurOtop Manual [2007] and Shore Protection Manual [1984] have been used for the design, construction and conservation of beaches etc. Apart from the two manuals mentioned, the guidelines for wave run-up and overtopping have been provided by Besley [1999] on EA Overtopping manual, the TAW [2002] Technical Report on Wave run-up and Overtopping at dikes and significant new informations from EC CLASH by several nations. EurOtop Manual advances all these informations in current practice, also extend the predictions on wave run-up and overtopping given in CIRIA/CUR Rock manual, Revetment manual, British Standards BS6349, the ISO TC98 and also US coastal Engineering Manual. It also includes case studies and sample calculations for different structures. Along with this manual, it also provides an online calculation tools to assist on overtopping predictions for different structures.

Problem I

An impermeable structure has a smooth slope of 1 in 2.5 and is subjected to a design significant wave, $H_s = 1.5$ m at a water depth, $d = 4.0$ m. Design mean wave period is $T_m = 6$ s. Water depth at structure toe at high water is $d_{toe} = 2.5$ m. (Assume no changes in the refraction coefficient.) Find the surf-similarly parameter (also called the Iribarren number) for the use of wave run-up and wave overtopping calculations for long-crested, irregular waves on impermeable (without water penetration) and permeable slopes. Find the height above the still-water (SWL) to which a new revetment must be built to prevent wave overtopping by the design wave.

Solution

The surf-similarly parameter for irregular waves depends on the wave steepness and structure slope.

6. Wave Run-up and Overtopping

Deep Water

First calculate the deep water, un-refracted wave height, H_0 from where measured back out to deep water. Using the depth where waves measured, and assuming $T = T_m = 6\,\text{s}$ and $H = H_s$ gives,

$$\frac{d}{L_0} = \frac{2\pi d}{gT^2} = \frac{2\pi(4.0)}{(9.81)(6)^2} = 0.0711$$

$$\frac{H}{H_0} = \sqrt{\frac{C_0}{C} \cdot \frac{1}{2n}} = K_s$$

$$\frac{H}{H_0} = 0.9694, \text{ the shoaling coefficient, } k_s$$

From Table C-1, of the shore Protection Manual [1984] (refer — Appendix A)

Therefore,

$$H_0 = \frac{1.5}{0.9694} = 1.54\,\text{m}$$

Toe of structure

Next, shoal the deepwater wave toe depth, $d = 2.5\,\text{m}$ at the toe of the structure.

From Table C-1, of the shore Protection Manual [1984] (refer — Appendix A) for

$$\frac{d_{toe}}{L_0} = \frac{2\pi d_{toe}}{gT^2} = \frac{2\pi(2.5\,\text{m})}{(9.81)(6)^2} = 0.0445$$

$$K_s = \frac{H_{toe}}{H_0} = 1.0417$$

$$H_{toe} = 1.54(1.0417) = 1.6\,\text{m}$$

Deepwater wave steepness, S_m

$$S_m = \frac{(H_s)_{toe}}{L_m} = \frac{2\pi H_{toe}}{gT_m^2} = \frac{2\pi(1.6)}{9.81(6)^2}$$

$$S_m = 0.028$$

Surf-similarity parameter, ξ_m

$$\xi_m = \frac{\tan\alpha}{\sqrt{S_m}} = \frac{1/2.5}{\sqrt{0.028}} = \frac{0.4}{0.1687}$$

Therefore,

$$\xi_m = 2.37$$

To prevent overtopping, based on Waal and van der Meer [1992],
As $\xi_m > 2.0$
Therefore,

$$\frac{R_{2\%}}{(H_s)_{toe}} = 3.0$$

Hence,

To prevent overtopping for impermeable slope, $R_{2\%} = 3.0(1.6) = 4.8$ m from the MSL to be provided for the revetment.

Whereas, in the case of permeable slopes,

Assuming double layered rubble mound slope,

when $\xi_m > 1.5$

$$\frac{R_{2\%}}{(H_s)_{toe}} = 1.17(\xi_m)^{0.46}$$
$$= 1.17(2.37)^{0.46} = 1.74$$

Therefore, $R_{2\%} = 1.635(1.6) = 2.78$ m from the MSL to be provided for the revetment.

References

Ahrens, J. P. (1988). Irregular wave run-up on smooth slopes, CETA No. 81-17, U.S. Army Corps of Engineers, Coastal Engineering Research Center, Ft. Belvoir, VA.

Ahrens, J. P., and Seelig, W. N. (1996). Wave run-up on beaches, *Proc. 25th Int. Conf. Coastal Engrg*, pp. 981–993.

Battjes, J. A. (1974). Surf similarity, *Proc. 14th Int. Conf. Coastal Engrg.*, vol. 1. pp. 466–480.

Besley, P. (1999). Overtopping of seawalls — design and assessment manual, R&D Technical Report W 178, Environment Agency, Bristol.

CIRIA/CUR. (1991) *Manual on the Use of Rock in Coastal and Shoreline Engineering*, CIRIA Special Publication 83, London, CUR Report 154, Gouda, The Netherlands.

Coastal Engineering Research Center (1984). Shore Protection Manual, U.S. Army Corps of Engineers, WES, Vicksburg.

CEM (Coastal Engineering Manual) (2002). U.S. Army Corps of Engineers, Engineer Manual 1110-2-1100, Washington, D.C. (6 volumes).

De Waal, J. P., and van der Meer, J. W. (1992). Wave runup and overtopping on coastal structures, *Proc. 23rd Int. Conf. Coastal Engrg.*, 2, pp. 1758–1771.

EurOtop. (2007). *Wave Overtopping of Sea Defences and Related Structures: Assessment Manual*, Environmental Agency, U.K. www.overtoppingmanual.com.

Hughes, S. A. (2003). Estimating irregular wave runup on smooth, impermeable slope, ERDC/CHL CETN-III-68, U.S. Army Engineer Research and Development Center, Vicksburg, MS.

Hughes, S. A. (2004). Wave momentum flux parameter for coastal structure design, Coastal Engineering, Elsevier, The Netherlands.

Hunt, I. A. (1959). Design of seawalls and breakwaters, *J. Waterways and Harbours Division*, 85 (WW3), 123–152.

Iribarren, C. R., and Nogales, C. (1949). Protection des ports, *XVIIth International Navigation Congress*, Section II, Communication. pp. 31–80.

Kobayashi, N. (1999). Wave runup and overtopping on beaches and coastal structures, *Advances in Coastal and Ocean Engrg.*, Vol. 5, World Scientific, Singapore, pp. 95–154.

Mase, H. and Iwagaki, Y. (1984). Runup of random waves on gentle slopes, *Proc. 19th Coastal Engrg. Conf.*, pp. 593–609.

Mase, H. (1989). Random wave runup height on gentle slope, *J. Waterways, Port, Coastal, and Ocean Engineering*, 115(5). ASCE, 649–661.

McConnell, K. J. (1998). Revetment systems against wave attack: a design manual, Thomas Telford, London.

Nielsen, P. and Hanslow, D. J. (1991). Wave runup distributions on natural beaches, *J. Coastal Res.* 7(4), Coastal Ed. and Res. Found., pp. 1139–1152.

Owen, M. W. (1980) *Design of Seawalls Allowing for Wave Overtopping*, Hydraulics Research Station, Wallingford, England. Report EX 924b.

Owen, M. W. and Steele, A. A. J. (1991). Effectiveness of recurved wave return walls, HR Wallingford, Report SR 261.

Reeve, D., Chadwick, A. and Fleming, C. (2012). *Coastal Engineering Processes*, Theory and Design Practice, Spon Press, Abington, UK.

Saville Jr., T. (1958). Wave run-up on composite slopes, *Proc. 6th Int. Coastal Engrg. Conf.*, pp. 691–699.

Sorensen, R. (1977). *Basic Coastal Engineering*, Springer, New York.

Stockdon, H. F., Holman, R. A., Howd, P. A. and Sallenger, A. H. (2006). Empirical parameterization of setup, swash, and runup, *Coastal Engrg.*, 53, pp. 573–588.

Van der Meer, J. W. and Stain, C. J. M. (1992). Wave runup on smooth and rock slopes of coastal structures, *J. Waterways, Port, Coastal and Ocean Engineering*, Vol. 118, No. 5, September/October 1992.

Van der Meer, J. W. (1998) *Rock Slopes and Gravel Beaches Under Wave Attack*. Ph.D. Dissertation, Delft Hydraulics Communication No. 396, Delft Hydraulics Laboratory, Emmeloord, The Netherlands.

TAW 2002. Technical Report — Wave run-up and wave overtopping at dikes, Technical Advisory Committee for Flood Defence in the Netherlands (TAW), Delft.

US Army Engineer Research and Development Center (1984), Vicksburg, MS, 39180, USA, Jeffrey.A.Melby@usace.army.mil.

Chapter 7

Scour Around Marine Structures

7.1 Introduction

Scour around structures is one of the major problems, often dictating their stability in the coastal and offshore environments. Scour is simply the removal of sediment particles in the vicinity of any obstruction due to the flow of water. Scour, for a layman, is simply the erosion of bed material due to a solid obstruction for the flow of water of ocean or a river. This phenomena can be experienced when one stands on the beach and is exposed to the return flow of waves that results in removal of sand around the feet. It is due to the interaction between an obstacles to the fluid flow field. As a matter of fact, all these processes interact with one another, which makes the process of scour very complex.

Scour occurs where, the sediments get eroded from the seabed under the influence of waves and currents. Scour nearby a structure can be attributed to the localized divergence in the sediment transport rate in the vicinity of the structure, thereby lowering the existing bed level. All the offshore structures require support from the ocean bed in the form of footings or raft or anchors or piles. When a foreign object is placed on the seabed, the flow pattern changes around the structure and erosion of the bed takes place. The scour will be more if the structure in an area of loose sand or when it is exposed to high velocity current. It can even cause the tilting of gravity structures. In the case of piles, it reduces the depth of fixity, which in turn, reduces the lateral load carrying capacity of the pile and the overburden pressure. Scour also causes problems in the case of structures like navigational structures and pipelines. Scour mainly depends on geometry of the structure, hydraulic parameters of flow and the sediment characteristics of seabed.

7.2 Mechanism of Scour

7.2.1 *General*

When structures are founded in potentially mobile sediments, it is important to quantify the degree of mobility and hence scouring of those sediments. Because a wide variety of installation types are placed in the marine environment, a general methodology for assessing the scour potential is required for each type. Considering the flow near an obstruction like pile, the dominant feature is the development of system of vortices around the pile. These vortices act as a vacuum cleaner and aid the erosion process at the cylinder. Some of the basic mechanisms responsible for the scour can be explained for a two-dimensional flow in the following manner. The flow boundary layer approaching a vertical cylinder generates a pressure gradient on the upstream face of the cylinder and the secondary flow results due to the interception of the incoming velocity by the obstruction as shown in Fig. 7.1. This has the effects of (i) formation of vortex (ii) downward reflection of the flow streamlines around the cylinder as shown in Fig. 7.2. The dimensions of the vortex system primarily depend on the shape and

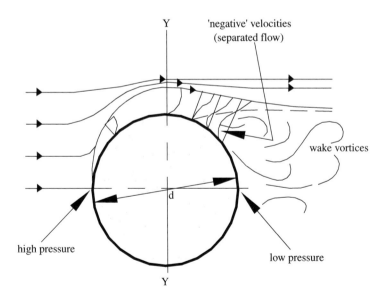

Fig. 7.1 Pressure gradient around the cylinder.

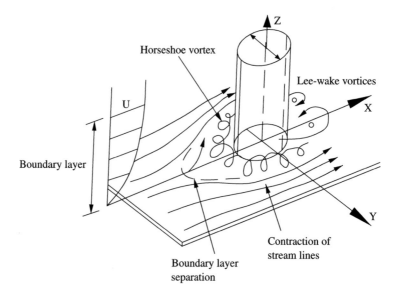

Fig. 7.2 Flow streamlines around the cylinder.

diameter of the obstruction. Flow velocity and sediment characteristics [Melville, 1975]. In cohesive sediments, the extent and depth of the scour are controlled by the size of the horseshoe vortex. Some sediment deposition may occur in the wake region of the flow. Erosion of cohesive soil occurs when the fluid shear is sufficient to overcome the tensile strength and the submerged weight of the bed materials.

7.2.2 *Fluid mechanism of scour*

The foregoing aspects give a base for an understanding of the complicated flow field in the vicinity of the obstruction. The processes that develops scour due to pure wave action are found to be different from those associated with steady current. Depending upon the environmental forces, the processes that develop scour are grouped into the following.

(i) Scour due to steady current
(ii) Scour due to the wave action
(iii) Scour due to simultaneous action of waves and current

7.2.3 Scour due to steady current

The principal flow feature and its situation can be described by considering a circular cylinder mounted normal to a plane surface. When the incoming flow velocity is intercepted by the seabed, the flow velocity decreases from a maximum at the free surface to zero at the seabed due to the presence of obstruction. This results in development of stagnation pressure ($\rho_w U^2/2$) that decreases with depth from the water surface. This pressure difference drives the flow and causes a strong vertical fluid jet, which descends along the upstream face of the obstruction (i.e. towards the decreasing velocity). A recirculating eddy (primary vortex) is formed when the flow impinges on the sea bed, the resulting vortex system wraps around the cylinder and trails off downstream. The horseshoe vortex, as shown in earlier figure, is indicated by the spiral curve wrapping around the cylindrical base. Flow patterns in the wake of the cylinder also influences the scour. Wake vortex systems are formed by the rolling up of the unstable shear layers generated at the surface of the pier and these are detached from both the sides of the pier at the separation line. They are shed alternatively from the pier and are trails off downstream as shown in Fig. 7.3 [Breusers and Raudkivi, 1991]. In the wake region, due to the pressure differences, the velocity gradient on the boundary layer causes the flow to occur in the wake, which move towards the bottom. Sediment particles are dislodged free along the front portion of the pile and carried away downstream by the

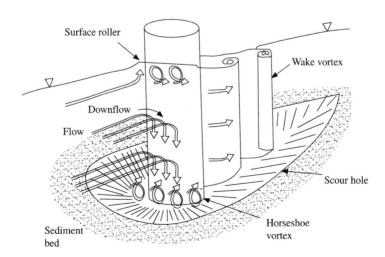

Fig. 7.3 Vortex formation [Breusers and Raudkivi, 1991].

vortex system due to combination of temporal agitation at the bed [Shen et al. 1966].

7.2.4 Scour due to waves

In the case of scour developed by waves only, the mechanism is different from that associated with steady flow. The steady current is thought of having persisted over a great distance so that the boundary layer (region of velocity gradient flow) is relatively thick. In contrast, the fluid particle travel for an oscillatory flow is not great enough for a substantial boundary layer height to develop. Therefore, the horseshoe vortex is smaller and weaker than that produces by steady current flow. Sediment scour in oscillatory flow appears to be caused by the acceleration of the primary flow past the obstruction, as well as in the variation of the flow patterns in the wake. Even when a wave induced scour hole fully developed the orbital current boundary layer remains thin and does not contribute to the formation of the horseshoe vortex.

7.2.5 Scour due to simultaneous action of waves and current

In a wave-current boundary layer flow, as the wave action increases, the mean current speed near the bed is progressively reduces whist the shear stress is increased. Thus, it appears that the ratio of wave to current speed as well as their magnitudes influences the scour development in wave-current flow. In the case of combined waves and current, the presence of waves modifies the vortex structure considerably. The near bed flow consists of formation of vortex and its shedding on the lee side of the pile (which is formed due to the interaction between the two shear layers emanating from the side edges of the pile). Each shed vortex lifts the grains into the upper portion of the vortex by the updraft. Entrainment of bottom sediments is proportional to the shear acting at the sediment water interface. If the steady flow velocity is stronger than the oscillating flow velocities, the equilibrium hole size and shape is expected to be greater than the one developed from steady current alone. This is because of stagnation pressure of the square of the velocities at two different heights. The implication is that, given an increased velocity due to both waves and currents, the secondary flow is stronger. Hence, the stronger the secondary flow, the stronger is the vortex system. Rate of scouring is expedited due to the superposition of waves on current. However, the size and pattern of scour hole tended to be of the same size and pattern as that due to the current alone.

7.3 Sediment Dynamics of Scour

The movement of bottom sediment involves, transport and erosion deposition. For the purpose of defining major physical differences in the dynamics of the sediments founded in marine environment, bottom sediments can be considered under two major groups viz.,

(i) **Cohesionless sediments** in which the main stabilizing force is acting and interlocking between the individual particles that act as discrete mass.
(ii) **Cohesive sediments** in which the primary stabilizing force is the physicochemical force between the grains and is stabilized in their position on the seabed by electrostatic forces between adjacent grain surfaces. The net attractive inter-particle surface forces which are electrochemical in nature and to some extent these forces impart the resistance to erosion.

Scour occurs when the fluid forces overcome the stabilizing forces. For cohesionless sediments, the magnitude of fluid shear stress needed to initiate grain movement can be determined from shields parameter [Keulegan and Carpenter, 1958]. But in the case of cohesive sediments, a generally accepted relationship has not yet been found. The bottom current velocities to develop the considerable fluid shear stress for the cohesionless soil is in the order of 0.20 m/s to 0.50 m/s and for cohesive soil (clay) it is 0.1 m/s to 2.2 m/s [Niedoroda and Swift, 1981]. In cohesionless soils, after the particles are dislodged, the particles move on or close to the sea floor (bed load transport) and cohesive sediments tend to get suspended and transported within the turbulence. Vanoni [1975] gave a chart for estimating the critical velocity to cause incipient motion of quartz sediment as a function of mean grain size (D_{50}). For mean grain size in the range of 0.2 mm to 100 mm, the critical velocity increases with increase in grain size.

7.4 Types of Scour

7.4.1 *General*

When there is an obstruction to the flow in the ocean, like an offshore structure, the bed in the immediate vicinity will be eroded, thereby depressions develop. This is called scour hole. Scour is generally classified into different types based on flow field, bed materials, and bed levels and also type of obstruction. The different types of scour are discussed in detail below.

7.4.2 General scour

General scour is usually defined as the bed lowering observed in rivers or channels as a response to increasing water discharge. General scour which occurs irrespective of the existence of the structure. Another aspect to take into account is the fact that general scour occurs in a river in its natural state, i.e., without the presence of structures, which in turn generate other types of scour, called local scour. Normally, the time scale for general scour is longer than the time scale for local scour.

7.4.3 Local scour

Local scour which occurs when water flows around obstructions in channels, generating complex flow patterns, increased flow velocities, turbulence and due to any local obstruction to flow in ocean e.g. step-sided scour pits around structures.

The intermittent structures tend to amplify the flow velocities and turbulence levels depending upon their geometry. It can also generate vortices and exert erosive forces on bed sediments, owing to which, the local enhancement of sediment movement rates and erosion are attained. The phenomenon of local scour around an obstruction is shown in Fig. 7.4. Local scour has received wide attention from researchers, mainly because it is easier to rig the model test set-up as well as to carry out the tests, in the

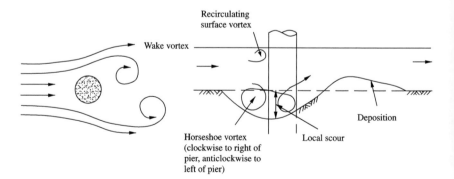

Fig. 7.4 Local scour.

event of which the time taken to simulate the local scour as per a model law will not require considerable time. Particular or unusual cases can also be model tested as needed.

Information on local scour is well known for the following features:

- bridge piers (of various shapes)
- bridge abutments and other similar structures
- river training works and linear revetments
- spur dikes (groynes) and other banks at various angles to the flow direction
- weirs and sills
- closures of cofferdams and diversion works.

In some cases, the extent, shape and surface gradients within the scour hole can also be predicted. The factors affecting the local scour is shown in Fig. 7.5.

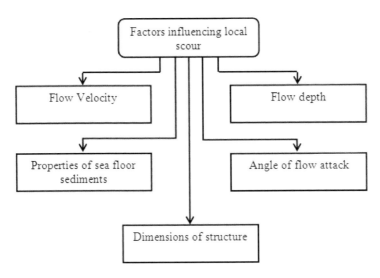

Fig. 7.5 Factors affecting local scour.

7.4.4 *Degradation scour*

It is generally referred to as lowering of bed levels along a river reach. This is often witnessed when the reach strives to regulate its gradient according to its corresponding flows and sediment loads the flow carries. When the sediment budget of the entry of sediment load is not compensated with the actual transport capacity, then degradation starts from upstream and proceeds to the downstream. In case the downstream posses a greater sediment transport capacity, degradation starts from the downstream end. From this description of the processes involved, it can be seen that, to obtain early warning of channel degradation problems at a structure, monitoring the changes over both upstream and downstream end for a significant distance is mandatory.

7.4.5 Boat scour

Scour may also increase due to the effects of navigation, in terms of the "blockage" or displacement effect of the vessel, causing increased local velocities around it, the currents caused by the vessel's propeller the surface waves (wake) caused by the vessel (which usually have most effect on the banks).

The effects of navigation on the amounts of scour may be considered by adding the associated flow velocities to those which occur in the absence of boats. In some cases the effects are likely to be minimal, due to the short durations and small velocities in relation to the natural velocities. In canals, the passage of boats is likely to be the principal cause of any scour.

7.4.6 High-head scour

Erosion of bedrock occurs in high-head situations, affecting massive hard igneous rocks as well as weaker sedimentary rocks. The primary mechanism for rock detachment in such situations is generally attributed to pressurisation of the joints.

7.4.7 Global or dishpan scour

Apart from the scour in the vicinity of the structure, changes in the sea bed due to the flow pattern can be witnessed over a wider region from the local scour. The combined local-global scour development is illustrated in Fig. 7.6. [Angus and Moore, 1982]. This type of scour is called global scour or dishpan scour resembles that of sand waves.

7. Scour Around Marine Structures   219

Fig. 7.6 Representation of global and local scour development around a jacket structure [Angus and Moore, 1982].

7.5 Scour Failures and Evolution

Ocean bed is composed of easily movable/replaceable materials, which when subjected to erosion in floods, leads to a fall in bed level. Scour in the vicinity of structures may be due to the combined effects of general, constriction and local scour, and may also be increased by the effects of navigation. When the scour level reaches the base of footing, the structure is at the risk of scour failure. Less intensity of scour will be experienced when the sub structure is subjected to lateral ground pressures and water forces. Scour adjacent to piled foundations may result in a loss of skin friction and load bearing capacity. Due to lateral loads and hydrodynamic forces, piles are subjected to unplanned bending stresses.

For local scour holes, four phases of evolution have been identified in model tests of clear-water scour [Hoffmans and Verheij, 1997]:

- initiation
- development
- stabilisation
- equilibrium

The rate of erosion is very high/rapid during the initiation phase, and the materials eroded from the upstream slope goes into suspension. The scour hole depth further aggravates in the development phase. The upper part of the upstream slope is in approximate equilibrium and development occurs beyond it, enlarging the hole while its shape remains approximately constant. The development of the hole results in a progressive decay in the velocity near the bed and rate of erosion. A near equilibrium state is said to be achieved during the stabilisation phase, when the erosion rate is small at the base of hole, although erosion continues at the more exposed and vulnerable position near the top of downstream slope, which results in lengthening of scour hole. In the equilibrium phase, the dimensions of the scour hole are virtually fixed. In the case of structures such as piers and piles, the scour hole usually originates along either sides of the structure and at the points of maximum width maximum depth of scour hole is observed, attributing to the high flow velocities. When the hole deepens further, they increase in extent and (unless the structure is extremely wide) join to form a single scour hole around the upstream end of the structure.

7.6 Scour Due to Vertical Walls

(i) Non-breaking waves

The maximum scour depth from the bed level (S_m) due to vertical walls because of the non-breaking waves can be predicted by Eq. (7.1) given below.

$$\frac{S_m}{U_{rms}T_p} = \frac{0.05}{[\sinh(k_p d)]^{0.35}} \qquad (7.1)$$

The root mean square of horizontal velocity $(U_{rms})_m$ can be established by knowing the significant wave height and wave number associated with the peak wave period.

$$\frac{U_{rms}}{gk_p T_p H_s} = \frac{\sqrt{2}}{4\pi \sinh(k_p d)} \left[0.54 \cosh\left(\frac{1.5 - k_p d}{2.8}\right) \right] \qquad (7.2)$$

where,

S_m = Maximum scour depth from bed level
g = gravity
d = water depth
T_p = peak wave period (period at which maximum energy occurs)
U_{rms} = Root mean square of horizontal velocity
k_p = Wave number associated with T_p
H_s = Significant wave height

7. Scour Around Marine Structures

Similarlly, the maximum scour depth from the bed level (S_m) due to non-breaking waves at the bed in the vicinity of the breakwater can be predicted by establishing the maximum orbital velocity.

$$S_m/D = 0.5[1 - e^{-0.175(KC-1)}] \qquad (7.3)$$

where,

$KC = (U_m T_P)/D$

and S_m = Maximum scour depth from bed level
D = Diameter of the circular head
T = Regular wave period
U_m = Maximum wave orbital velocity at bed
KC = Keulegan – Carpenter number

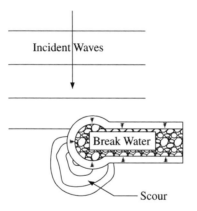

Fig. 7.7 Scour due to breakwater.

(ii) Breaking waves

When the amplitude of wave reaches a critical level and suddenly all the energy contained in the wave transformed into turbulent kinetic energy resulting in immense force acting on the structure. In such a case, where the velocity of the particle is high removes the sediment particles from the sea bed and scour occurs. Scour due to breaking waves can be predicted by Eq. (7.4) given below.

$$\frac{S_m}{(H_s)} = \sqrt{(22.72\frac{d}{(L_p)_o} + 0.25)} \qquad (7.4)$$

The above equation is valid when $\frac{d}{(L_p)_o}$ is 0.011 to 0.045 and $\frac{(H_s)}{(L_p)_o}$ is 0.015 to 0.04.

Whereas the maximum scour at sloping structure due to breaking waves can be predicted by the Eq. (7.5) given below.

$$\frac{S_m}{H_s} = 0.01 \left(\frac{T_p\sqrt{gH_s}}{d}\right)^{3/2} \qquad (7.5)$$

where,

L_p = Significant wave length
H_s = Significant wave height

7.7 Pipelines

7.7.1 *General*

Deploying offshore pipelines as a means to transport oil and gas has gained popularity over the recent years. Stability of the pipeline is essential and it is receiving increasing attention. Three situations can be distinguished. (1) Pipelines crossing areas where, ships might anchor in cases of emergency. (2) Pipelines crossing fishing areas. (3) Pipelines in areas where no interference from human sources is to be expected. In situation either the pipeline must be buried so deep that anchors cannot reach it or there must be a cover layer which gives perfect protection against anchors. It is doubtful whether this latter goal can ever be achieved economically, in view of the large mass of the anchors of large vessels. In cases, where, the pipelines must be buried, two techniques can be used; (a) burying by jetting or fluidization; (b) dredging a trench and covering the pipelines after it has been placed in the trench. The different possible configurations for submarine pipelines are illustrated in Fig. 7.8.

7.7.2 *Scour around pipelines*

Numerous empirically defined independent parameters governs the development of scour holes beneath the pipelines. The parameters that describe the boundary conditions for local scour processes are the depth averaged flow velocity, pipe diameter, D, clearance of the pipe from the seabed, y; water depth, d, and grain diameter, D_g. Additional parameter under wave action are wave height, H, and wave period, T. From these primary parameters other parameters can be derived, such as bottom roughness

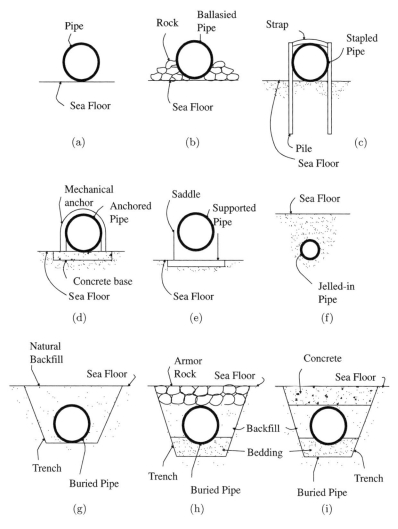

Fig. 7.8 Examples of possible configuration for submarine pipelines.

orbital velocities, and ultimate sediment transport capacities, S. In general, pipeline exposure will decrease significantly when the relative height, y/D ($y = e + 0.5D$), of the pipe above the seabed becomes and finally stop when y/D reaches values of -0.5 to -0.7. The scour underneath the pipe reaches a maximum if a pipeline lies on the seabed and $e = 0$ [Leeuwestein et al. 1985].

7.7.3 Scour around pipelines due to current action

Pipelines which are laid on the seabed will in general, be exposed to the lower part of the current profile. The pipeline thus protrudes a certain height into this profile and determines the flow pattern around it, resulting in the erosion and sedimentation of bottom material.

The sudden increase in transport capacity of the water passing through the pipeline can be a reason for scour under a pipeline. Sedimentation/settling of sediments occurs when there is a decrease in transport capacity. These gradients in transport capacities exist only temporarily until the bottom has changed into its equilibrium configuration according to the instantaneous flow conditions. The transport capacity will be denoted by "S" and subscripts "o" and "p" will be generally used to signify locations beyond the influence of the pipe and underneath the pipe, respectively. Although bed transport and suspended transport certainly should be treated separately, unless stated explicitly total transport will be used here for simplicity. When an equilibrium situation has been established, both the depth of the scour hole, y_s, and the gap, y, between the pipe and the bottom of the scour hole remain constant. The transport capacities satisfy the equality.

$$S_o = S_p \tag{1}$$

When a scour hole is developing underneath a pipe, there is no balance of transport capacities:

$$S_p > S_o \tag{2}$$

A more detailed description of the erosion phase distinguishes two cases; (1) no transport upstream of the pipe ($S_{ou} = 0$), and transport only underneath and possibly just downstream of the pipe (S_p, $S_{od} \# 0$); (2) transport at all locations ($S_p > S_o > 0$). In the second case there is a supply of sediment to the developing scour hole, with a possibility of exchange of sediment between the suspended and the bottom layer. Prediction of the rate of erosion and the ultimate scour depth on the basis of a sediment balance now seems to be possible, given a particular-flow pattern around the pipe.

Leeuwestein and Bijker [1985] identified three basic forms of erosion around a submarine pipeline:

1. Luff erosion, which occurs at the upstream side of the pipe and is caused by an eddy formation upstream of the pipe.
2. Lee erosion, which occurs at downstream side of the pipe and is caused by reemergence of the main flow over and the turbulent wake downstream of the pipe.

3. Tunnel erosion, which occurs under the pipe and is a direct consequence of the increased velocity underneath the pipe, compared with the undisturbed velocity.

Mao [1986] reported the formation of three types of vortices around a submarine pipeline resting on a plane bed. As illustrated in Fig. 7.9, one of the vortices A, formed at the nose of the pipe, the other two vortices, B and C, formed downstream of the pipe. It was also cited pressure coefficient distribution near a pipe in unidirectional flow measured by Bearman and Zdravkovich [1978] and deduced that the pressure difference between the upstream and downstream sides of the pipe may cause seepage underneath the pipe. The onset of scour according to Mao [1986], due to the combined action of the vortices and underflow, which leads to the formation of a small opening under the pipe as more and more sand particles are carried away. Figure 7.10 clearly illustrates the process of the onset of scour.

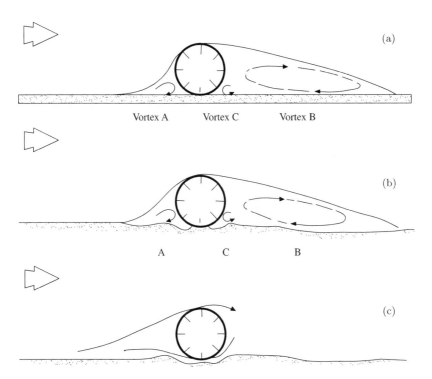

Fig. 7.9 Three-vortex system and onset of scour.

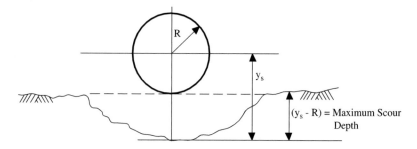

Fig. 7.10 Graphic illustration of the problem of scour around pipeline.

7.7.4 *Scour due to waves*

When scour is driven by the effect of waves alone, the erosion mechanism is unlike those associated with steady flow is relatively thick when compared to oscillatory flow in which fluid particle will not travel great distance thereby, substantial boundary layer will not be developed. Therefore, vortex will be smaller and weaker than that produced by steady current. Therefore, scour produced will also be smaller excepting only the ripple formation.

7.7.5 *Scour due to waves and currents*

In the case of ocean environment, scour takes place due to the combined action of waves and currents. For whatever be the situation, either weak current plus large wave velocity or strong current plus small wave velocity, in the points on the stagnation line a pressure gradient is developed, which is proportional to the difference in squares of instantaneous approach velocity.

Owing to the influence of both waves and currents, the resultant flow velocity is increased, due to the increased magnitude of flow velocity the secondary flow inclines to be much stronger, and hence stronger will be the horse shoe vortex system. Waves and current interaction can be discussed in three ways.

1. Large current and low wave velocity in the deep-water region: when the wave motion is in the same direction as the current the velocity field and secondary flows are enhanced. When the wave motion is reversed, phenomenon also reverses.
2. Low current and high wave velocity: in this case horseshoe vortex will form only when the current and wave directions are similar.

3. Equal current and wave velocity: the combined velocity varies between the combined value of current plus wave velocities and zero velocity. This situation is difficult to evaluate.

7.8 Maximum Scour Depth

An analytical method for estimating the maximum scour depth under the offshore pipelines due to currents was developed by Chao and Hennessy [1972]. This method provides an order of magnitude estimation of the possible scour hole depth.

The subsurface current is assumed to flow perpendicular to the longitudinal axis of the pipeline. Based on two-dimensional potential flow theory and the assumptions outlined by Chao and Hennessy, the discharge through the scour hole is,

$$q = U_0 \left(y_s - \frac{R^2}{2y_s - R} \right) \; for \; y_s \geq R, \tag{7.6}$$

where,

U_0 = undisturbed subsurface current over the top of the pipe
$R = D/2$ = diameter of the pipe
y_s = scour hole depth from the center of the pipe

The average jet velocity is

$$U_{avg} = \frac{q}{(y_s - R)} = U_0 \left[\frac{2(y_s/R)^2 - (y_s/R) - 1}{2(y_s/R)^2 - 3(y_s/R) + 1} \right] \; for \; y_s \geq R \tag{7.7}$$

If the velocity in the scour hole is greater than the free stream velocity, erosion may take place. The limit of scour is presumably reached when, because of the enlargement of the scour section, the velocity along the boundary has decreased to the point at which the boundary shear stress τ_b becomes equal to the critical tractive stress τ_{cr} of the sediment composing the erodible beds. The tractive stress τ_{cr} for a given sand grain size as suggested by Herbich [1981] is plotted in Fig. 7.11.

The boundary shear stress in the eroded channel is computed based on assumptions stated by Chao and Hennessy [1972]. The friction factor f_f, is estimated from the Reynolds number relationship reported by Lovera and Kennedy [1969], by using a Reynolds number R_e defined as,

$$R_e = \frac{U_{avg}(y_s - R)}{\gamma} \tag{7.8}$$

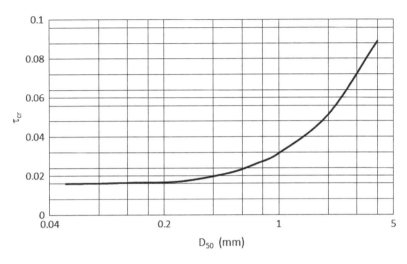

Fig. 7.11 Critical tractive stress versus grain size.

where γ is the kinematic viscosity of sea water. The roughness parameter is defined as $(R_h/D_{50}) \times 10^{-2}$, where R_h is the hydraulic radius, which is approximated, as $(y_s - R)$. Once the friction factor is known, the boundary shear stress is calculated by using the relationship of [Streeter, 1971] as:

$$\tau_b = \frac{\rho(U_{avg})^2}{8} \times f_f \qquad (7.9)$$

where ρ is the density of sea water. The maximum scour depth under the pipeline can then be determined by matching the calculated boundary shear stress with the critical tractive stress of the size of the sand grains composing the ocean floor.

7.9 Scour Protection

7.9.1 *General*

Scour often leads to instability of a structure, the essential part of structural stability relies on its toe stability. The failure of toe sources the entire structure to fail. The magnitude of any scouring as a result of structural influences is difficult to predict. It may sometimes be unobserved because maximum scour occurs during when the storm reaches the maximum height. Protection of toe mound is very crucial to assure the safety of the structure and also prevent sliding failure. It therefore needs to be provided to an adequate depth and be of sufficient size/stability to avoid the common failure modes. The location of the structure in relation to the wave breakpoint, form of structure with respect to reflectivity and nature of the seabed are crucial in deciding the nature of toe protection required. The action of scour is often catalyzed with changes in alignment, downdrift of groins, channels and structural roundheads. The purpose of providing flexible toe is to ensure that the foundation materials remains in its place beneath the structure until the depth of maximum limit of scour.

The methods available for the protection of scour and also to increase the stability of the structure are, (1) rip rap rock fill (2) protective mattress and (3) buried toe (4) sand bags or grout filled bags (5) concrete grout and (6) soil and structural improvement.

7.9.2 *Rip rap rock fill*

Deploying rip raps is the most convenient and cost effective methods of scour protection. This method involves fixing of apron or collar around the base of the structure in a rip-rap manner which serves as a protective layer.

7.9.3 *Protective mattress*

Scour protection by providing mattress is often used for pipelines, but it can also be embraced for other sea bed structures. Because of its flexible nature and installation in controlled manner, protection by mattress has an advantage over rock fill. There are different types of mattress such as fascine mattress, block mattress, cell mattress, concrete mattress and ballast mattress. Soil pins or anchors are provided to hold the mattress in its place. The soil pins or anchors has to be designed to resist the uplift and drag due to hydrodynamic forces. Filters made up of geo-textiles or geo-synthetic materials can also be used to prevent erosion underneath the hydraulic structures.

7.9.4 *Buried toe*

Where construction conditions permit, the cover layer is extended by burying the toe in an excavated trench to the depth of predicted scour. It may be appropriate to backfill the trench with granular fill or rock, depending on natural conditions.

7.9.5 Sand bags or grout filled bags

Stabilisation of scour holes is usually done using sand bags which require sufficient time for their installation at the desired location. Such measures are effective only for a shorter duration since they are likely to be undermined. Sand bags or grout filled bags are also used to underpin pipeline free spans, but may then require a protective covering of gravel or mattresses to provide protection to the underpinning material and to stabilise the bed around the pipeline [Van Dijk, 1980], prior to the installation of a protective layer of rock or mattresses.

7.9.6 Concrete grout

Voids, e.g. below footings, can be filled with concrete grout. Where this is carried out underwater the grout can be injected using tremie pipes. Formwork is needed to contain the grout, which can take the form of concrete bags or sheet piles, as described above, as well as conventional formwork. Grouting is carried out from the bottom up. As grouting progresses the tremie pipe should remain embedded in the freshly placed grout. Formwork should be watertight to prevent loss of grout and held down by dowels or similar to prevent uplift generated by pressure of the grout. The top of the formwork should be protected to avoid the loss of fines and cement, but vented to indicate when the formwork is full. A typical arrangement for the top of the formwork may consist of a highly permeable fabric held down with wire mesh, in turn held down with plywood.

7.9.7 Structural improvements

Laying deep foundations beyond the limit of maximum expected scour is one of the most sought after and effective structural solution. Structures including sills, drop structures etc. are often built in the close vicinity of bridges. The effect of contraction scour can be mitigated by erecting new spans facilitating additional flow area and thereby, reducing the flow velocity.

7.10 Bed Shear Stress

7.10.1 Bed shear stress due to waves

Sediment is mobilized under combinations of wave-current flow when the value of maximum shear stress, $\tau_{\max}$ exceeds τ_{cr}. (τ_{cr} is threshold value

7. Scour Around Marine Structures

of shear stress by which sediment particles begin to move across the seabed).

The mobility of the sediment can be quantified by comparing the shear stress calculated according to below formulae with the critical shear stress, defined by

$$\tau_{cr} = \theta_{cr} g (\rho_s - \rho) D_{50} \qquad (7.10)$$

where θ_{cr} is the critical shields parameter, g is the acceleration due to gravity, ρ_s and ρ are the densities of sediment and water, and D_{50} is the median grain diameter.

Note: θ_{cr} is obtained using Shield's curve for D^* (as discussed in Sec. 2.15) given by

$$D^* = D_{50} \left[\frac{(s-1)g}{v^2} \right]^{1/3} \qquad (7.11)$$

where $s = \frac{\rho_s}{\rho}$ and v is the kinematic viscosity of fluid (m²s⁻¹).

The wave related bed shear stress can be calculated from [Soulsby, 1997]

$$\tau_w = 0.5 \rho f_w u_{\max(-d)}^2 \qquad (7.12)$$

in which f_w is the wave friction factor, a function of the ratio A/z_0 where A is the amplitude of the orbital wave motion at the bed ($u_{\max(-d)} T/2\pi$), and z_0 is hydraulic roughness length defined by,

$$z_0 = D_{50}/12 \qquad (7.13)$$

For monochromatic waves, using small amplitude linear wave theory, U_w is determined as:

$$u_{\max(-d)} = \frac{\pi H}{T} \frac{1}{\sinh(kd)} \qquad (7.14)$$

where H is the wave height, T the wave period, k the wave number ($k = 2\pi/L$, L = wave length) and d the water depth.

The expression for f_w in rough turbulent flow is:

$$f_w = 1.39(A/z_0)^{-0.52} \qquad (7.15)$$

7.10.2 Current related bed shear stress

The grain related bed shear stress due to steady current τ_{cr} (Nm⁻²) is calculated from the expression [Soulsby, 1997]

$$\tau_{cr} = \rho C_D \overline{U}^2 \qquad (7.16)$$

where, ρ is the water density (kgm^{-3}), C_D the drag coefficient (a function of the ratio of water depth and hydraulic roughness length (z_0)) related to $\overline{U}$ the depth averaged current speed. $\overline{U}$ occurs at a height above the bed equal to 32% of the water depth and the tidal current speed at the surface is equal to 1.07.

$$\overline{U} \cdot C_D = \left[\frac{0.40}{\ln(d/z_0) - 1}\right]^2 \quad (7.17)$$

in which d is the mean water depth (m) and z_0 the hydraulic roughness length (m) relating to the bed sediment, which is defined in earlier section Eq. (7.13)

$$z_0 = \frac{k_s}{30} = \frac{2.5 D_{50}}{30} \quad (7.18)$$

where D_{50} is the median diameter (m) of the sediment. In this method, as in those below, a single representative grain size other than D_{50} can be used if appropriate.

7.10.3 Combined wave and current shear stress [Soulsby, 1995 and 1997]

The following approach for calculating the maximum bed shear stress $\tau_{\max}$(Nm^{-2}) due to the interaction of waves and currents has been adopted. Firstly the mean shear stress $\tau_{\max}(Nm^{-2})$ due to combined waves and currents is calculated:

$$\frac{\tau_m}{\tau_{cr}} = 1 + 1.2 \left(\frac{\tau_w}{\tau_{cr} + \tau_w}\right)^{3.2} \quad (7.19)$$

where τ_w is calculated from Eq. (7.12) and τ_{cr} from Eq. (7.16). This expression accounts for the non-linearities introduced when waves and currents interact and has been calibrated against laboratory and field data of Soulsby [1995, 1997]. The maximum shear stress is calculated by vector addition of τ_m and τ_w to give the magnitude of the shear stress vector [Ockenden and Soulsby, 1994], Fig. 7.12.

$$\tau_{\max} = \left[(\tau_m + \tau_w \cos\varphi)^2 + (\tau_w \sin\varphi)^2\right]^{0.5} \quad (7.20)$$

where φ is the angle (degree) between the wave and current shear stresses, or velocities to a reasonable approximation. When it is known, the angle between waves and currents should be used in Eq. (7.20), but an average value of φ equal to 45° can be assumed in place of a known value.

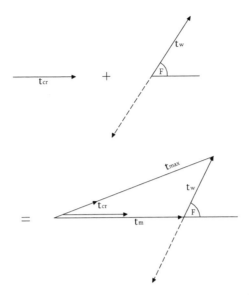

Fig. 7.12 Definition of shear stress vectors for wave, current and combined shear stresses.

References

Angus, N. M. and Moore, R. L. (1982). Scour repair methods in the southern North Sea, *Proc. 14th Annual Offshore Technology Conference*, OTC, Houston, Texas, May 3–6, 1982, paper no. 4410, 385–399.

Bearman, P. and Zdravkovich, M. (1978). Flow around circular cylinder near a plane boundary, *J. Fluid Mech.*, 89, 33–47.

Breusers, H. N. C. and Raudkivi, A. J. (1991). Scouring: Hydraulic Structures Design Manual Series.

Chao, J. L. and Hennessy, P. V. (1972). Local scour under ocean outfall pipelines, *J. Water Pollution Control Fed.*, 7, 1443–1447.

Hoffmans, G. J. C. H. and Verheij, H. J. (1997). *Scour Manual*, A. A. Balkema Publishers, Rotterdam.

Herbich, J. B. (1981). Scour around pipelines and other objects, in *Offshore Pipeline Design Elements*, Marcel Dekker, New York, xvi +233p.

Keulegan, G. H. and Carpenter, L. H. (1958). Forces on cylinders and plates in an oscillating fluid, *Journal of Research of the National Bureau of Standards*, 60(5), 423–440.

Leeuwestein, W. and Bijker, E. W. (1985). *The Natural Self-burial of Submarine Pipelines, Behaviour of Offshore Structures*, Elsevier, pp. 717–728.

Lovera, F. and Kennedy, J. F. (1969). Friction-factors for flat-bed flows in sand channels, *J. Hydr. Div., ASCE* Vol. 95, pp. 1227–1234.

Mao, Y. (1986). The Interaction Between a Pipeline and an Erodible Bed, Lyngby DTH Licentiatafhandling.

Melville. (1975). *Sedimentary Structures, Their Character and Physical Basis*, Vol. 2.

Melville, B. W. (1992). Local scour at bridge abutments, *Journal of Hydraulic Engineering*, 118(4), pp. 615–631.

Niedoroda, A. W. and Swift, D. J. P. (1981). Maintenance of the shore faces by wave orbital currents and mean flow.

Ockenden, M. C. and Soulsby, R. L. (1994). Sediment transport by currents plus irregular waves, Report SR 376, HR Wallingford, Wallingford, UK.

Soulsby, R. L. (1995). Bed shear stresses due to combines waves and currents, Section 4.5 in *Advances in Coastal Morphodynamcs*, eds. Stive, M. J. F. et al., Delft Hydraulics, Delft.

Soulsby, R. L. (1997). *Dynamics of Marine Sands, A Manual for Practical Applications*, Thomas Telford, London.

Streeter, V. L. (1971). *Fluid Mechanics*, McGraw-Hill, New York, 755p.

Vanoni, V. A. (ed). (1975). *Sedimentation Engineering*. Manuals and Reports on Engineering Practice, no. 54, ASCE, New York, USA.

Van Dijk, R. (1980). Experience of scour in the southern North Sea, Proc. Society for Underwater Technology One-day seminar on scour prevention techniques around offshore structures, 3–10 December.

Chapter 8

Design of Coastal Structures

8.1 Introduction

The structures in the coastal region in general said to be coastal structures which are an important component in any coastal development scheme including its protection. This chapter, provides an up-to-date technical advances in the design and construction of coastal structures. The need for coastal protection is mainly to provide beach and shoreline stability control by controlling erosion, to develop harbours, to stabilise the navigation channels at inlets through training walls, to protect the sea water intakes and outfalls and to retain or rebuild the natural systems (cliffs, dunes) or to protect buildings landward of the shoreline from coastal flooding.

For the design of these coastal structures, it is necessary to consider the physical factors such as geomorphology, material characteristics, tides, winds, storms, waves, currents, shoreline details, bathymetry, littoral drift and the availability of stones of different sizes, availability of appropriate equipment and the purpose for which the structure are to be constructed.

Design of coastal structures is often focused on design values for certain parameters, like the $p_{\max,2\%}$ or $p_{\max}$ for a design impact pressure, $Ru_{2\%}$ for a wave run-up level and q as mean overtopping discharge or $V_{\max}$ as maximum overtopping wave volume. A structure can then be designed using the proper partial safety factors, or through probabilistic approach.

Wave forces are the key forces for design of any coastal structures and are highly variable and depend on both the wave conditions and the type of structure being considered. When the structure is placed such that the incident waves are unbroken, then a standing wave will exist seaward of the wall and only the static and dynamic forces exist. These can be readily determined from linear wave theory, Dean and Dalrymple (1991). However, more commonly the structure will need to resist the impact of breaking or broken waves. The most widely used formula for estimating the static and dynamic forces in this situation is based on Goda (see Burcharth in Abbott and Price (1994) for a review). However, high localized shock forces may also arise due to breaking waves trapping pockets of air, which are rapidly compressed. The study of this phenomenon is an ongoing area of research and currently there are no reliable formulae for the prediction of such forces. The works of Allsop *et al.* (1996) and Bullock *et al.* (2000) provide some recent results. Several breakwaters are constructed using large blocks of rock (the "armour units") placed randomly over suitable filter layers. At locations, wherein large natural rocks is scanty, they are being replaced by numerous shapes of massive concrete blocks (e.g., dolos, tetrapods, CoreLocs, accropods, etc.) size of which depends on several inter-related factors, like, wave height, armour unit type and density, structure slope and permeability. Traditionally, the Hudson formula that expresses the weight of the armour unit proportional to H^3 has been widely used. This was derived from an analysis of a comprehensive series of physical model tests on breakwaters with relatively permeable cores and using regular waves. More recently (1985–1988) these equations have been superseded by the formula of Van der Meer (2011) for rock breakwaters.

8.2 Non-breaking Wave Forces

A vertical impermeable wall obstructs the kinetic energy in the waves a major portion of which is reflected and some spent in the wave run-up over it. The upward component of the energy over the wall can result in the wave crests to rise to double the incident wave in deeper waters. The downward component causes severe erosion and scour that would lead to the instability of the structure if proper toe protection is not provided.

Several analytical and laboratory and field investigations have been undertaken to develop formulae for predicting the dynamic pressure on walls due to waves. However, most of the formulae are based on monochromatic regular wave of constant height and period. Critical cases such as non-overtopping vertical wall, overtopping vertical wall, vertical wall with rubble foundation and non-overtopping vertical wall with different water depth on both sides are briefly discussed in this chapter.

Miche-Rundgren (1958)

Miche-Rundgren method of wave force estimation is based on the modified Sainflou's formula (1928). Non-breaking waves incident on smooth, impermeable vertical walls are completely reflected by the wall with a reflection coefficient of 1.0. Vertical walls built on rubble bases commonly adopted as toe protection or in the case of composite breakwaters will experience lesser reflection coefficient and hence reduced forces. For complete details on the wave forces on wall type structures one can also refer to CEM (2002).

The wave conditions at a structure and seaward of a structure (where reflection of waves are not accounted or negligible are shown) are given in Figs. 8.1(a) and 8.1(b).

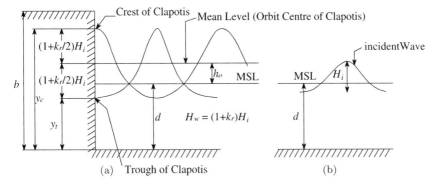

Fig. 8.1 Definition of term: non-breaking waves.

Definition of terms non-breaking wave forces are given below

d: water depth, H_i: height of incident wave, K_r: reflection coefficient, h_o: height of standing orbit center (mean water level at wall above SWL),

y_c: depth from clapotis crest $= d + h_o + \{(1+K_r)/2\}H_i$
y_t: depth from clapotis trough $= d + h_o - \{(1+K_r)/2\}H_i$
b: height of wall and wave height at the wall $= H_w = (1+K_r)H_i$

If $H_r = H_i$ then $K_r = 1$ and height of the clapotis or standing waves at the structure will be $2H_i$ and the height of y_c and y_t are given above. Any value of K_r less than 0.9 are not recommended for designing purpose. The resulting total hydrodynamic load when the wave trough is at the vertical wall is less than the hydrostatic loading if waves were not present and the water was at rest. The pressure distributions at the crest and trough at a vertical wall are shown in Figs. 8.2(a) and 8.2(b).

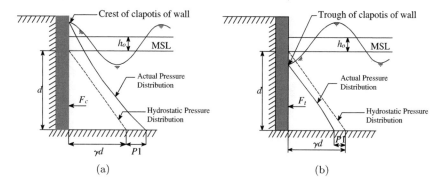

Fig. 8.2 Pressure distribution at crest and trough.

When the crest is at the wall, pressure increases from zero at the free water surface to $\gamma d + p_1$ at the bottom.

$$p_1 = \frac{1+K_r}{2}\left[\frac{\gamma H_i}{\cosh kd}\right] \tag{8.1}$$

When the trough is at the wall, the conservative values of pressure increases from zero at the water surface to $\gamma d - p_1$ at the bottom.

Pressure distribution for critical condition

The critical condition is with the differential water depth on both sides of the vertical wall with the crest on the deeper water side and the trough on the shallower water depth side as shown in Fig. 8.3. The pressure forces and moments are then calculated separately on either side of the wall.

The pressure, force and moment on the crest side are considered to be p_1, F_c and M_c. Similarly, on the trough side is considered to be as p_2, F_t and M_t, respectively. The net pressure is based on the difference between crest and trough side, i.e. $p_1 - p_2$. Similarly, force and moment are $F_t - F_c$ and $M_t - M_c$, respectively. Figures 8.4 to 8.6 permit the prediction of the forces and moments resulting from a non-breaking wave on a wall.

The forces and moments do not include the forces and moments due to hydrostatic pressure at SWL. From the above figures, we can get F_w and M_w which are the forces and moments due to the wave alone. If hydrostatic

8. Design of Coastal Structures 239

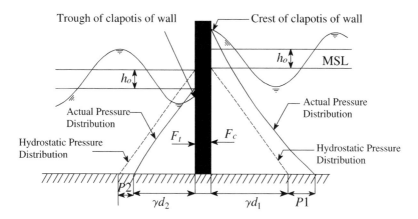

Fig. 8.3 Pressure distribution at critical condition.

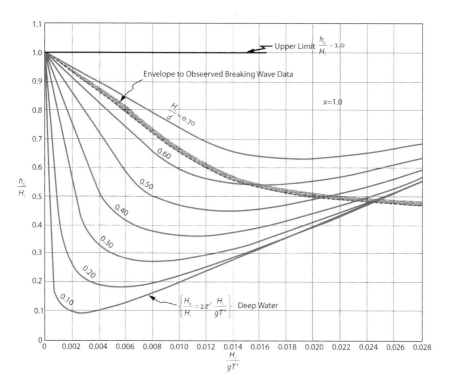

Fig. 8.4 Non-breaking waves; $K_r = 1.0$.

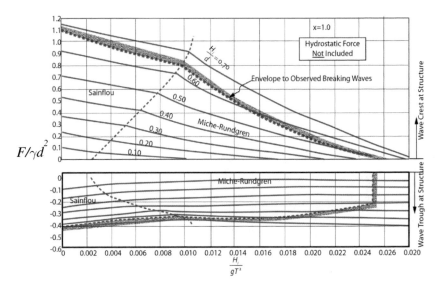

Fig. 8.5 Non-breaking wave forces; $K_r = 1.0$.

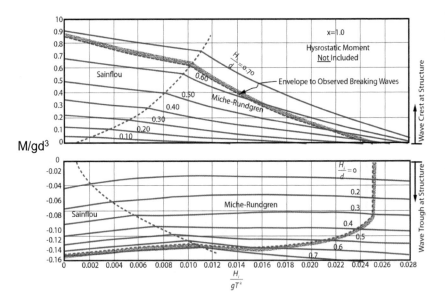

Fig. 8.6 Non-breaking wave moment; $K_r = 1.0$.

effects (e.g. seawalls), the total force, F_t and moment, M_t are to be found by the expressions

$$F_t = \frac{\gamma d^2}{2} + F_w \qquad (8.2)$$

$$M_t = \frac{\gamma d^3}{6} + M_w \qquad (8.3)$$

8.3 Wave Forces on Walls and Rubble Mound Structures

The rubble mound and vertical types are the common coastal structures. The evaluation of wave forces on such structures also important.

8.3.1 *Rubble mound structures*

In the case of a rubble mound breakwater, like groins or breakwaters, the size of the individual stone/armor or artificial block forming the primary or the cover layer should be designed such that it is stable when exposed to the action of high waves even during severe sea state. The forces acting on a rubble stone on a slope α is schematically represented in Fig. 8.7. Equating the different forces, we get

$$mg \sin \alpha + (\text{force due to waves}) = \mu mg \cos \alpha \qquad (8.4)$$

wherein "m" is mass of stone or the armour block, μ is a friction coefficient. The diameter of stone is D_g, the density is ρ_s and ρ_w are the density of water, respectively. The wave force on the armour, F_a can be expressed as

$$F_a = 0.5 f_w \rho_w u^2 \cdot D_g^2 \cdot \text{constant} \qquad (8.5)$$

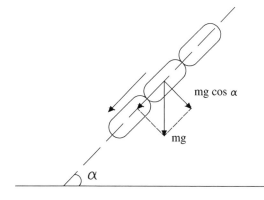

Fig. 8.7 Forces on an armour unit in waves — definition sketch.

Herein, f_w is the friction factor and on assuming a long wave acts on the armour, "u" can be $u =$ constant, $(gH_b)^{1/2}$. Hence, the design mass of stone M can be calculated as,

$$M = \frac{\rho_s H_b^3}{K_D(\rho_s/\rho_w - 1)^3(\mu\cos\alpha - \sin\alpha)^3} \quad (8.6)$$

The breaking wave height, H_b can be assumed to be the design wave height, K_D as a constant that takes into account the characteristics of the armor layer and ρ_s and ρ_w are the densities of armour layer for e.g. rock and sea water, respectively. However, the formula of Hudson (1959) is being widely adopted for obtaining the weight of the armour block which is given as

$$W = \frac{\gamma_s H_D^3}{K_D(S_r - 1)^3 \cot\theta} \quad (8.7)$$

where H_D is the design wave height in "m", γ_s is the unit weight of the armour unit, θ is the angle of revetment with the horizontal, submerged weight of the armour, $S_r = (\rho_s/\rho_w - 1)$. K_D is a dimensionless stability coefficient, determined through physical modelling. This value for K_D vary for different kinds of armour blocks, type of placing the armour units, number of layers, slope of the surface and nature of waves whether breaking or non-breaking. $K_D =$ around 2 to 4 for natural quarry rock, the lower value for breaking waves, whereas, for tetrapods, that is being widely used it is 7 and 8 for breaking and non-breaking waves, respectively. For other artificial interlocking concrete blocks, like CoreLocs, accropods it is about 10 or even more. Refer CEM (2006) for more details.

8.3.2 Pressure distribution on an overtopped wall

Pressure distribution on an overtopped wall is brought out in Fig. 8.8. Wave overtopping of vertical walls provides a reduction in the total force and moment because the pressure above its crest will be absent. The effect of overtopping should be considered in the design as this would exert seaward pressure on the wall caused by saturated backfill or ponding water. The total pressure is $= \gamma d + p_1$, as p_1 in the case of overtopping type vertical wall can be established by incorporating the reflection coefficient and shown in relation below.

$$p_1 = \frac{1 + K_r}{2}\left(\frac{\gamma H_i}{\cosh 2kd}\right) \quad (8.8)$$

8. Design of Coastal Structures   243

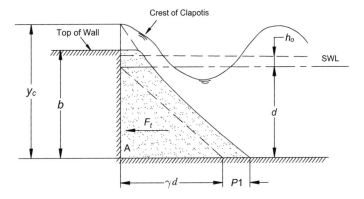

Fig. 8.8 Pressure distribution on an overtopped wall.

8.3.3 Minikin's method for a wall on a low rubble mound

The design of any structure to provide a non-overtopping condition by the design wave is often not feasible in economic perspective. The pressure distribution for this case is shown in Fig. 8.3 already. We see overtopping taking place and if it is not too severe, the truncated pressure distribution results in a force F^1 which is proportional to F, and where F is the total force that would act against the wall if it extended up to the crest of the clapoitis and F is determined from Fig. 8.5. The relationship between F' and F is,

$$F' = r_f F \qquad (8.9)$$

where r_f is a force reduction factor due to lower height of wall given by

$$r_f = \begin{cases} \dfrac{b}{y}\left(2 - \dfrac{b}{y}\right) & \text{when } 0.5 < \dfrac{b}{y} < 1.0 \\ 1.0 & \text{when } b/y \geq 1.0 \end{cases} \qquad (8.10)$$

where b is defined in Fig. 8.8 and y can take y_c or y_t. The relationship between r_f and b/y is shown in Fig. 8.9. Similarly, the reduced moment about point A is given by

$$M' = r_m M \qquad (8.11)$$

where the moment reduction factor, r_m is given by

$$r_m = \left(\frac{b}{y}\right)^2 \left(3 - 2\frac{b}{y}\right) \text{ when } 0.5 < \frac{b}{y} < 1.0$$

$$r_m = 1.0 \text{ when } \frac{b}{y} \geq 1.0 \tag{8.12}$$

These equations are valid when either the wave crest or wave trough is at the structure, provided the correct value of y is used.

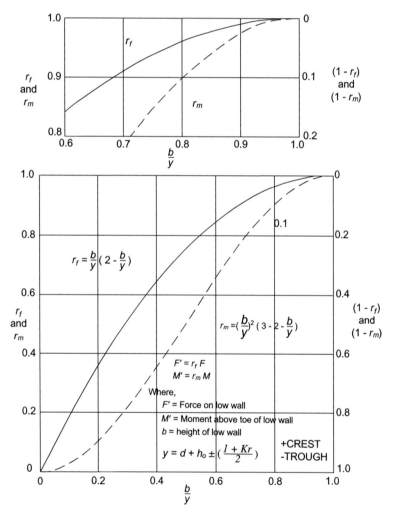

Fig. 8.9 Reduction factor.

8.3.4 Wall of rubble foundation

The pressure distribution for this case is shown in Fig. 8.10.

$$\left.\begin{array}{r}F'' = (1 - r_f)F \\ M_A'' = (1 - r_m)M\end{array}\right\} \qquad (8.13)$$

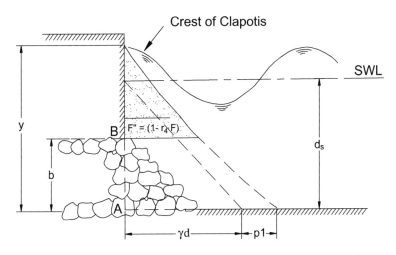

Fig. 8.10 Pressure distribution on the wall over rubble mound.

F'' is force against the wall above the foundation. M_A'' is moment about the bottom (point A on Fig. 8.10). Moment at point B is given by

$$\left.\begin{array}{l}M_B'' = (1 - r_m)M - b(1 - r_f)F \\ \text{or} \\ M_B'' = M_A'' - bF''\end{array}\right\} \qquad (8.14)$$

The values of $(1 - r_m)$ and $(1 - r_f)$ may be obtained from Fig. 8.9.

8.3.5 Breaking wave forces on vertical walls

Minikin Method (1955, 1963) can give forces 15 to 18 times those calculated for non-breaking waves. The maximum pressure assumed to act at the SWL is given by,

$$p_m = 101\gamma \frac{H_b}{L_D} \frac{d_s}{D}(D + d_s) \qquad (8.15)$$

p_m: maximum dynamic pressure, H_b: breaker wave height, d_s: is the depth at the toe of the wall, D: is the depth of one wavelength in front of the wall and L_D: is one wavelength in water of depth D.

The pressure decreases parabolically from p_m at SWL to zero at a distance of $H_b/2$ above and below the SWL (Fig. 8.11).

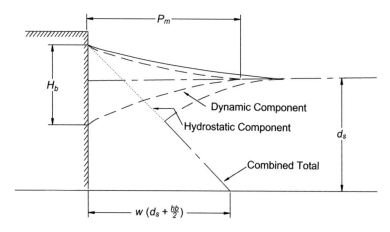

Fig. 8.11 Minikin wave pressure diagram.

The force represented by the area under the dynamic pressure distribution is

$$R_m = \frac{p_m H_b}{3} \qquad (8.16)$$

i.e. the force resulting from dynamic component of pressure and overturning moment about the toe is

$$M_m = R_m d_s = \frac{p_m d_s H_b}{3} \qquad (8.17)$$

Minikin's method was originally derived for composite breakwaters. D and L_D are the depth and wavelength at the toe of the substructure and d_s is the depth at the toe of the vertical wall (i.e. the distance from SWL down to the crest of the rubble substructure). For caissons and other vertical structures where no substructure is present, the formula has been adapted by using the depth at the structure toe as d_s, while D and L_D are depth and wavelength a distance one wavelength seaward of the structure. Hence,

$$D = d_s + L_d m \qquad (8.18)$$

where L_d is the wavelength where the water depth is equal to d_s, m is the nearshore slope. The triangular hydrostatic pressure distribution is shown in Fig. 8.11, the pressure is zero at the breaker crest (taken at $H_b/2$ above SWL), and increases linearly to $\gamma(d_s + H_b/2)$ at the toe of the wall. The total breaking wave force on the wall per unit length is

$$R_t = R_m + \frac{\gamma\left(d_s + \frac{H_b}{2}\right)^2}{2} = R_m + R_s \qquad (8.19)$$

where R_s is the hydrostatic component of breaking wave on a wall, and the total moment about the toe is,

$$M_t = M_m + \frac{\gamma\left(d_s + \frac{H_b}{2}\right)^3}{6} = M_m + M_s \qquad (8.20)$$

where M_s is hydrostatic moment.

8.3.6 Wall on a rubble mound

The procedure for calculating forces and moments is similar to that outlined in the example presented at the end of this chapter, except that the ratio d_s/D, is used instead of the nearshore slope when using Fig. 8.12.

8.3.7 Wall of low height

When the top of a structure is lower than the crest of the design breaker, the dynamic and hydrostatic components of wave force moments can be corrected by using Figs. 8.13 and 8.14. Figure 8.13 is a Minikin reduction factor to be applied to the dynamic component of the breaker wave force equation

$$R'_m = r_m R_m \qquad (8.21)$$

Figure 8.14 gives a moment reduction factor "a" for use in equation,

$$M'_m = d_s R_m - (d_s + a)(1 - r_m) R_m \qquad (8.22)$$

or

$$M'_m = R_m[r_m(d_s + a) - a] \qquad (8.23)$$

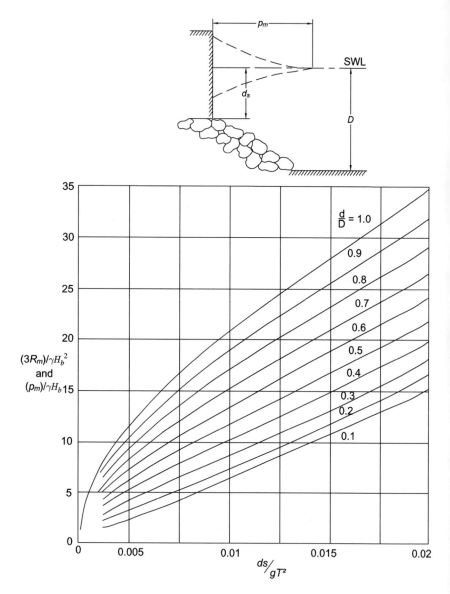

Fig. 8.12 Dimensionless Minikin wave pressure and force.

8. Design of Coastal Structures

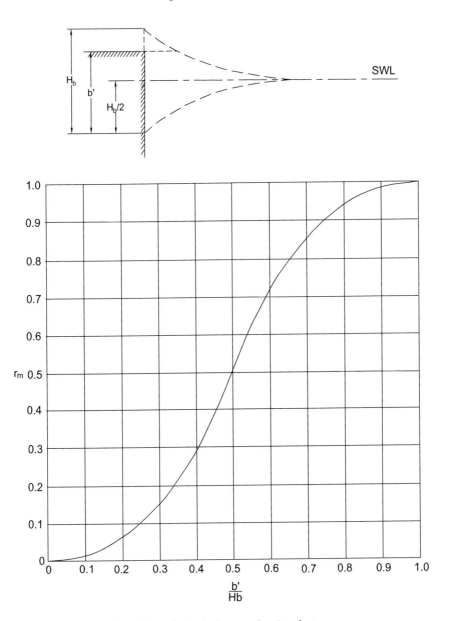

Fig. 8.13 Minikin force reduction factor.

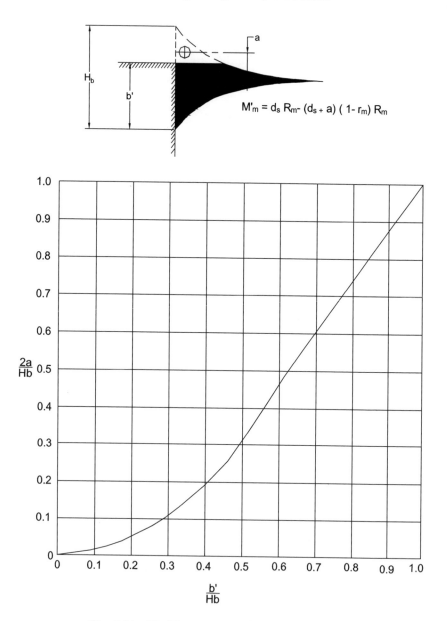

Fig. 8.14 Minikin moment reduction for low wall.

8. Design of Coastal Structures

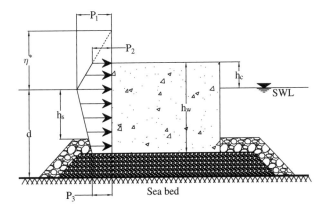

Fig. 8.15 Definition sketch of Goda's wave induced pressure under crest.

8.4 Goda's Method of Breaking Wave Force (1974) (CEM Table VI-5-53, p. VI-5-139)

It is based on model tests and design against the single largest wave force in design sea state is considered. It also uses the highest wave in a wave group. The design wave height, H_{design} is estimated at a distance of $5 \times H_s$ seaward of breakwater ($H_{design} = 1.8H_s$). It is later modified to incorporate random wave breaking model and assumes trapezoidal shape for pressure distribution along front wall of caisson imbedded into the rubble mound. The uplift pressure distribution is assumed to be triangular.

H_{max} should be based on Goda's random wave breaking model, and $H_{max} = 1.8H_s$ is recommended. Elevation to which wave pressure is exerted,

$$\eta* = 0.75(1 + \cos\beta)H_{max} \qquad (8.24)$$

in which, β = direction of waves with respect to breakwater normal (for waves approaching normal to breakwater, $\beta = 0$)

Pressure on front of vertical wall,

$$p_1 = 0.5(1 + \cos\beta)(\alpha_1 + \alpha_2 \cos^2\beta)\gamma H_{max}$$

$$p_2 = \begin{cases} \left(1 - \dfrac{h_c}{\eta*}\right)p_1 & \text{for } \eta* > h_c \\ 0 & \text{for } \eta* \leq h_c \end{cases}$$

$$p_3 = \alpha_3 p_1$$

Effect of wave period on pressure distribution

$$\alpha_1 = 0.6 + 0.5 \left(\frac{2kd}{\sinh 2kd}\right)^2$$

minimum = 0.6 (deep water), maximum = 1.1 (shallow). Increase in wave pressure due to shallow mound,

$$\alpha_2 = \text{minimum of } \frac{h_b - h_s}{3h_b}\left(\frac{H_{\max}}{h_s}\right)^2 \text{ or } \frac{2h_s}{H_{\max}}$$

Linear pressure distribution

$$\alpha_3 = 1 - \frac{h_w - h_c}{d}\left(1 - \frac{1}{\cosh kd}\right)$$

h_b = water depth at $5H_s$ seaward of breakwater

Buoyancy and uplift pressure,

$$p_u = 0.5(1 + \cos\beta)\alpha_1\alpha_3\gamma H_{\max}$$

(Japanese found that $p_u = p_3$ was too conservative.)
Decrease in pressure from hydrostatic under wave trough

$$p = \begin{cases} \gamma z & : -0.5H_{\max} \leq z < 0 \\ -0.5\gamma H_{\max} & : z < -0.5H_{\max} \end{cases} \qquad (8.25)$$

There are different types of structures in the coastal area. However, most of the structures predominantly falls under these three categories.

- Rubble mound structures
- Vertical and composite structures
- Retaining structures
- Marine piled structures

The design principles and considerations of these four types of structures are discussed in the following sections. The typical design principles of rubble mound breakwaters are given in a comprehensive manner.

8.5 Rubble Mound Structures

8.5.1 *General*

The rubble mounds are gravity structures which derive their stability largely from the weight of the armouring units which cover and protect the core. The entire structure is typically graded in layers from the large exterior stone or concrete units through two or more layers of intermediate sized materials to small quarry run sizes at the core and finer material beneath it.

8.5.2 Armour layer

The primary purpose is to resist as well as to dissipate incident wave energy, in order to protect the inner layer from severe wave attack and therefore creating tranquil conditions on its lee side. The size of the armour units for rubble mound structure section is calculated by using the Hudson formula, which is recommended by CERC (1984). The stability number (N_s) given in Eq. (8.26).

$$N_s = \frac{H_{des}}{\Delta_a D_a} = (K_D \cot\theta)^{1/3} \quad (8.26)$$

$$D_a = D_{60} = \left(\frac{M_a}{\rho_a}\right)^{1/3} \quad (8.27)$$

$$\frac{H_{des}^3}{\Delta_a^3 D_a^3} = K_D \cot\theta \quad (8.28)$$

$$\rho_s D_a^3 = \frac{\rho_s H_{des}^3}{\Delta_a^3 K_D \cot\theta} \quad (8.29)$$

$$W = \frac{W_r H_{des}^3}{K_D (S_r - 1)^3 \cot\theta} \quad (8.30)$$

where W = weight of an individual armour unit in the primary cover layer, W_r = unit weight of armour unit, H_{des} = design wave height at the structure site in meters, S_r = specific weight of armour unit relating to water at the structure, $S_r = (W_r/W_w)$, $\Delta_a = (S_r - 1)$, W_w = unit weight of seawater = 1025 kg/m^3, θ = angle of structure slope measured with the horizontal in degrees, K_D = stability coefficient, (for TETRAPOD and KOLOS in breaking wave condition the K_D is 4.5 for random placement).

It is important to mention that the Hudson formula is not valid for slope lesser than 1.5. The Stability Number according to Van der Meer (1988) for different breaking waves are given below.

For plunging breakers

$$N_s = 6.2 P_b^{0.18} \left[\frac{S_d}{\sqrt{N_w}}\right]^{0.2} \frac{1}{\sqrt{\xi_n}} \quad (8.31)$$

For surging breakers

$$N_s = 6.2 P_b^{0.18} \left[\frac{S_d}{\sqrt{N_w}} \right]^{0.2} \sqrt{\cot \theta} \, \xi_m^{P_b} \qquad (8.32)$$

N_w is the number of waves, P_b an overall porosity of breakwater, $P_b = 0.1$ for armour layer over an impermeable layer, $P_b = 0.4$ for armour over a filter layer over a coarse core, $P_b = 0.6$ for structures built entirely over armour stone, S_d is the damage level and ξ_m is the surf similarity parameter.

8.5.3 *Underlayer*

The purpose of underlayer is to support armour units. It is designed so that armour units should not penetrate through the voids of the under layer.

$$D_{15} \text{ (armour)} < 5 \, D_{85} \text{ (underlayer)}$$

The size of stone in under layer will be considered from $W/10$ to $W/15$ (as per CERC, 1984). Rough angular quarry stones are considered for underlayer in most cases for which $W_r = 2650$ kg/m³.

8.5.4 *Core layer*

The purpose of under layer is to support armour units and underlayer. It is designed so that the underlayer units should not penetrate through the voids of the core. It prevents the penetration of the waves and sediment. It avoids the sediment passing through it.

$$D_{15}(\text{underlayer}) < 5 D_{85}(\text{core}).$$

The size of stone in core layer is taken as $W/100$ to $W/225$ (as per CERC, 1984). The rough angular quarry stones are considered for core layer in most cases for which $W_r = 2650$ kg/m³.

8.5.5 *Toe mound*

It provides stability against scouring and sliding. Its principle function is to support the armour stones as well as scour protection.

The size of stone in toe mound is taken as $W/10$ to $W/15$ (as per CERC, 1984). The rough angular quarry stones are considered for the toe layer in most cases for which $W_r = 2650$ kg/m³. The thumb rule to adopt the width and the thickness of the toe berm is $3D$ and $2D$, respectively.

8.5.6 Thickness of armour and underlayer

The thickness of the armour and underlayer is calculated by following equation.

$$t = nK_\Delta \left(\frac{W}{W_r}\right)^{1/3} \quad (8.33)$$

where, n is the number of stones ($n = 3$ is minimum), K_Δ is the layer coefficient, W is the weight of the armour unit in primary cover, W_r is the mass density of armour unit, representative diameter,

$$D = \left(\frac{W}{W_r}\right)^{1/3}$$

8.5.7 Crest elevation

Crest width for the breakwater is 3 to 5 times of the armour block. The crest elevation of the breakwater is fixed from data such as storm surge, run up, mean highest high water level.

Crest elevation = R + free board + Design Water Level

where, R = wave run up estimated as per CEM (2006).

Free board may be adopted in calculating the design elevation to give free height for exceptional cases of storms and cyclone waves that hit the toe of the structure to avoid dangers. Typically, a free board of 0.5 m will be adopted in most cases. However, to avoid inundation due to long period swells, the crest elevation of armour layer will be further elevated so that the proposed elevation is sustainable for extreme conditions as well.

8.5.8 Filter layer

It separates seabed from the entire structure.

$$D_{15}(\text{core}) < 5D_{85}(\text{filter layer}).$$

The filter layer is recommended for a thickness of 300 mm with 10 mm to 50 kg rubble stones following the suggestions of SPM of the order of $W/2000$ to $W/6000$. A typical design of breakwater is given in Problem VI of this chapter.

8.6 Vertical and Composite Structures

The design of vertical and composite structures is based on the identification of the main failure processes and then to compute the characteristic dimension of the selected structure type to ensure that the principal loadings remain below the structure's resistance when suitably factored. In the design of related such structures, the main emphasis has historically been on balancing the horizontal (and perhaps up-lift) forces against caisson weight hence the friction forces.

These are mainly mass concrete/masonry structures with the primary purpose of maintaining tranquility behind the structure. The vertical wall structure is designed so that there is no tension in the structure. The dead weight of the structure resists all the external loads. Typical vertical and composite structure configurations are shown in Fig. 8.16.

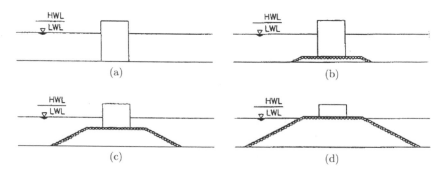

Fig. 8.16   Vertical and vertically composite structure configurations.

Vertical wall and composite type structures have been built since the earliest development of coastal zone. The primary purposes of these structures are to protect areas of water for navigation, anchorages or sheltered moorings and to protect working areas within and around the harbours or also to defend the coast against erosion. The composite type are taking particular account of local conditions and/or materials. As the name implies, the mound level of high-mound composite breakwater is higher than the low water level (LWL). For high mound breakwater, the waves tend to break at the mound, whereas, low mound breakwater does not cause wave breaking on the mound as it is submerged. This leads to instability of high-mound composite due to impulsive pressure and scouring triggered by breaking waves. For this reason the low mound composite breakwater is quite commonly considered for construction.

For the purpose of reducing the reflection of waves and the force generated by breaking wave on the vertical wall, concrete caisson blocks are positioned in front of it. This is commonly referred to as composite breakwaters, which is now called the horizontally composite breakwater. This type of structures are conventional, since the critical damages in a vertical wall breakwaters are often strengthened by concrete armour blocks or rubbles by dissipating the wave energy. Modern horizontally composite breakwaters employ shape-designed concrete blocks such as tetrapods. The design begins with the study of tension at the base of the structure as shown in the below equation.

$$f = \frac{W}{A} \pm \frac{M}{Z} \quad (8.34)$$

For zero tension $f = 0$.

$$\frac{W}{b} > \frac{W \times e}{\frac{1 \times b^2}{6}} \quad \therefore e < \frac{b}{6}$$

where, e is the eccentricity of the load, b is the width of the caisson, W is the dead load of the structure, M is the moment due to the external force especially wave force and Z is the section modulus of the base.

Check for sliding and overturning

The most commonly addressed failure mode of monolithic structures is sliding backwards against direct wave action. This primarily depends on the horizontal loads however influenced by the uplift force. The design of a breakwater's upright caisson section must be stable against sliding and overturning (Fig. 8.17), and to accomplish this, safety factors against sliding and overturning must be greater than 1.2. In most cases, sliding is more severe than overturning, especially when the breakwater crown is relatively low.

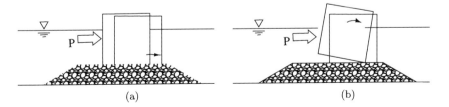

Fig. 8.17 Sliding and overturning of upright section.

Check for sliding

$$\mu \frac{R_v}{R_o} > C_s \qquad (8.35)$$

where μ is the coefficient for sliding, R_v is the total vertical load, R_o is the total horizontal load and C_s is the factor of safety against sliding (normally taken as 1.4).

To assess the sliding and overturning stability of the upright section, the weight (W), buoyancy (B), the horizontal wave induced force (F_h) and uplift force (U) must be considered. The dynamic uplift pressure is assumed to vary linearly from the seaside to the lee side.

Safety Factors $(S.F.)$ are calculated as follows: For sliding, the friction due to the net downward forces opposes the horizontal wave induced force,

$$S.F. = \mu(W - B - U)/F_h \qquad (8.36)$$

Check for overturning:

$$\frac{M_s}{M_r} > C_r \qquad (8.37)$$

where M_s is the total resisting sectional moment due to dead load, M_r is the total horizontal moment, C_r is the factor of safety against overturning (normally taken as 1.5).

For overturning, moments are calculated about the lee side toe,

$$S.F. = (M_W - M_u)/M_p \qquad (8.38)$$

for a symmetric section with no eccentricity:

$$M_W = 0.5\beta(W - B)$$
$$M_U = b_u U = 2/3\beta U$$
$$M_p = d_h F_h$$

Stability of the block masonry wall

The stability of the block masonry wall can be determined using the extended Goda pressure formula, and it should be examined at each level of the blocks, namely, the stability of all the blocks above each level should be examined. The pressure existing between blocks, which acts as an uplift pressure on the upper block, can be assumed to be equal to the horizontal pressure occurring at the same level (Tanimoto and Ojima, 1983).

It should be noted that the friction coefficient between flat concrete is 0.5, and that any interlocking effect between the blocks should be considered if present.

8.7 Retaining Structures

The structures that engineered to retain a mass of sand and/or rubbles are generally referred to as retaining structures. These structures are also frequently used to house variations in grade, increases in right-of-way and provides reinforcement to the toe of slopes. Retaining structures can be classified in a broad sense conferring to their inclination of face when greater than 70°. Whereas, slopes with inclination flatter than 70° are usually considered as retaining walls. Apart from the classifications, there are several types of retaining structures which includes the gravity structures, sheet piles, cantilever structures, and anchored earth. Retaining structures withstand loads from the horizontal stresses in the ground which is considered to be as earth pressure and also from the anchorages that are used to support the structures. The pressure on the wall relies on whether the structure is moving away or towards the earth and also on the characteristics of the soil. The calculation of earth pressure, shear force and bending moment for the design of retaining structures can be found in Atkinson (2007) or Punmia (2005) and IS 4651: part 2 (1989).

8.7.1 Gravity retaining walls

The stability of gravity retaining structures are based on the self-weight which it use to resist or hold the earth pressures. They are typically constructed from materials of heavyweight, mostly concrete and rubbles. Such heavy material structures are designed to deliver significant resistance against several failure kinds which includes seismic activities, overturning, sliding and enhances the bearing capacity. Crib wall can be referred to a common type of retaining wall. It is commonly used in highway cuts in the coastal region and to support roadway because of its cost effective construction, workability and for time saving. Another example of gravity retaining structures is gabions. It is used as a coastal protection structure for controlling erosion and also scour protection along the waterways and river banks. They are normally made of wire mesh carriers that are packed with rubbles of relatively uniform sizes and arranged on top of each other to form a retaining wall.

In the last few decades, Mechanically Stabilized Earth (MSE) walls are well developed and widely recognized as gravity retaining walls. In a MSE wall, an area or part of a zone is well reinforced and it delivers retention resistance. Wire mesh, welded bars, steel strips and bar mat are some of the metallic reinforcing elements used for reinforcement. Reinforcements are also provided with polymeric reinforcement like geo-grids. Depending on the site condition woven or non-woven geosynthetics are used to retain the soil from the action of waves.

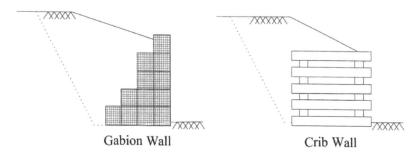

Gabion Wall　　　　　　　　Crib Wall

Fig. 8.18　Gravity walls.

8.7.2　Sheet pile walls

Sheet piles are usually a steel sheet that can be hammered into the ground to support the earth pressures. Based on the site condition the sheet piles are either driven by impact hammer or gently pushed into the ground. In some cases, vibratory equipment is required for installing sheet piles. In order to support a deeper excavations, tie-back anchors are required to strengthen the sheet piles especially in the location of conjunction. The design of sheet pile will be based on numerous forces acting on sheet pile wall. Typically the active earth pressure on the back of the wall and the passive earth pressure on the front of the wall are estimated to arrive at the section. The active pressure tends to move the wall away from the backfill whereas, the passive earth pressure resists the wall movement.

The main considerations for the design of sheet piles are structural safety and stability, demands of aesthetics, availability of material, equipment and expertise, constructability and ease of maintenance and also its durability. Detailed design principles and consideration can be found in *Design of Sheet Pile Walls* manual (1994).

8. Design of Coastal Structures

Often the sheet piles are connected or partially connected sheet piles in order to construct continuous walls for waterfront structures which includes a small waterfront boat launching facilities to a large dock. Typical cross-section of sheet pile is shown in Fig. 8.19. One of the major advantages of sheet piles over other retaining structures is that dewatering in the location is not required compared to other retaining structures.

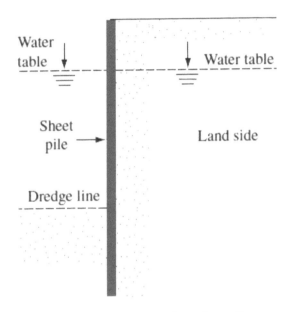

Fig. 8.19 Typical waterfront sheet pile.

Several types of sheet piles are commonly used in construction: (a) wooden sheet piles, (b) precast concrete sheet piles, and (c) steel sheet piles. Aluminium sheet piles are also marketed. Wooden sheet piles are used only for temporary, light structures that are above the water table. The most common types are ordinary wooden planks and Wakefield piles. The wooden planks are about in cross section and are driven edge to edge. Wakefield piles are made by nailing three planks together, with the middle plank offset by 50 mm to 75 mm. Wooden planks can also be milled to form tongue-and-groove piles. The selection of steel sheet piles is based on the section modulus and moment of inertia. Typical sheet pile section and its corresponding section modulus and moment of inertia is tabulated in Table 8.1.

Table 8.1. Typical sheet-pile section and its section modulus and moment of inertia.

Sketch of section	Section modulus m^3/m	Moment of inertia m^4/m
409 mm; 12.7 mm; 15.2 mm; Driving distance = 500 mm	326.4×10^{-5}	670.5×10^{-6}
379 mm; 12.7 mm; 15.2 mm; Driving distance = 575 mm	260.5×10^{-5}	493.4×10^{-6}
304.8 mm; 9.53 mm; 9.53 mm; Driving distance = 457.2 mm	162.3×10^{-5}	251.5×10^{-6}
228.6 mm; 9.53 mm; 9.53 mm; Driving distance = 558.8 mm	97×10^{-5}	115.2×10^{-6}

(*Continued*)

8. Design of Coastal Structures 263

Table 8.1. (*Continued*)

Sketch of section	Section modulus m^3/m	Moment of inertia m^4/m
12.7 mm, Driving distance = 500 mm	10.8×10^{-5}	4.41×10^{-6}
9.53 mm, Driving distance = 406.4 mm	12.8×10^{-5}	5.63×10^{-6}

8.7.3 Anchored earth structures

The resistance to earth pressure is achieved by anchoring a soil nail or tie back to the dense layer or rocky layer beneath the loose sand. Such technique of strengthening is referred to as anchored and reinforced earth structures. It is done by the assembly of facing units that are tied to rods or strips that are held in place by friction. The resistance of the ties to movement is controlled by the portion of the anchors/nails that are located behind the theoretical active wedge also known as the failure wedge.

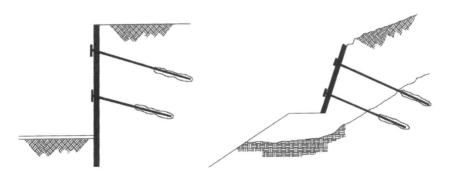

Fig. 8.20 Anchored earth structures.

8.8 Marine Piled Structures

For most of the marine structures, a strong base of the structure can be achieved by only pile foundation. Piling secures the foundation of structures as it delivers the essential support required for any kind of marine structure to be built. Piling is the procedure of setting deep foundations into the ground, usually using wood, steel or concrete. Reinforced Cement Concrete (RCC) marine piles are basically elongated cantilever slender piles installed in loose soil/clay and/or rocky bed. The fixity depth i.e. the effective depth below the sea bed or dredge level should be approximated based on the site and load condition in order to begin the construction. In general, the marine piles are considered to be as fixed end cantilever members and the depth of fixity is calculated as per equivalent length producing equal deflection. Provisions of Indian Standard code of Practice of Piles (IS 2911) are generally followed. On the other hand, the fixity depth can be established by using the modulus of sub-grade reaction and applying spring support to the pile member of the jetty or wharf structure. The pile capacity is usually designed by considering the soil capacity, and the structural design is carried over based on the shear and moment forces. It is to be renowned that in most practical cases the percentage of reinforcement is ruled by serviceability criteria of crack width, which is as per the provision of IS or BS codes. In general, marine piles are heavily reinforced. In the case of steel piles, a similar method is practiced however, the crack width consideration will be neglected as it is not required in steel piles. Steel piles can take large deflections and especially for mooring and breasting dolphins piles it helps in design.

Piling is also a cost-effective way of construction for marine structures like jetties, berthing structures and pipe trestles. Fenders in front of the berth either connected or unconnected to the berth usually protect and reduces the berthing forces imposed by the ships and in turn protects the pile rows. During berthing, the impact induces a great impact force whereas, the resistance offered by the vertical pile to lateral loading is trivial. Hence the deck should be designed to support by a combination of vertical and raking piles. The piles in the berthing head of a cargo jetty are required to carry the following loadings: Lateral loads from berthing forces, Lateral loads from the pull of mooring ropes, Lateral loads from wave forces on the piles, Current drag on the piles and moored ships, Lateral loads from wind forces on the berthing head, moored ships, stacked cargo, and cargo handling facilities, Compressive loads from the dead weight of the structure, cargo handling equipment, and from imposed loading on the deck slab, Compressive and uplift forces induced by overturning movements due to loads 1 to 5 above and in some parts of the world piles may also have to carry vertical and lateral loads from floating ice, and loading from earthquakes.

It is possible that the combination of above forces may occur whereas the cumulative of all the forces may not happen. Hence, the design can be carried over by considering the combination of forces as well. For instance, the wind, wave, and current forces can occur simultaneously and in the same direction, the forces due to berthing impact and mooring rope pull occur in opposite directions. Taking the case of a vertical pile acting as a simple cantilever from the point of virtual fixity below the sea bed, and receiving a blow from the ship with a force H applied at a point A, the distance moved by the point A can then be calculated equation shown below,

$$\text{distance moved}, y = \frac{H(e+z_f)^3}{3EI}.$$

The bending moment M, on the pile is equal to $H(e+z_f)$ therefore,

$$\text{Work done} = \frac{M^2(e+z_f)}{6EI}.$$

In the case of a pile fixed against rotation by the deck slab of a structure, then the distance moved can be estimated by,

$$\text{distance } y \text{ moved point } A = \frac{H(e+z_f)^3}{12EI}.$$

The bending moment caused by a load at the fixed head of a pile to $= \frac{1}{2}H(e+z_f)$, and thus the work done is the same as for single pile. In the case of a piled wharf erected parallel to a sloping shore line, the piles supporting the rear of the deck, being more deeply embedded than those at the front will resist a much higher proportion of the horizontal forces imposed on the fendering. It may be necessary to consider sleeving the rearward piles to equalize the flexural resistance. If the rear of the deck is abutting a retaining wall such as a sheet pile wall, virtually the whole of the horizontal forces on the deck will be transmitted to the wall.

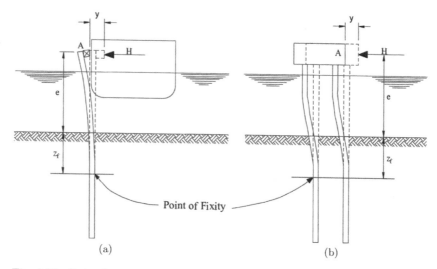

Fig. 8.21 Lateral movement of fender piles due to impact force from berthing ship (a) Single free-headed pile (b) Group of fixed-headed piles.

8.9 Problems

Problem I

Estimate the wave force and moment on the vertical wall of height 4.5 m in a water depth of 3 m due to the incident wave height of 1.5 m and the wave period of 6 sec or 10 sec.

Solution

Given: Wall height, $b = 4.5$ m; water depth, $d = 3$ m; wave height, $H_i = 1.5$ m; wave period, $T = 6$ sec or 10 sec.

For $T = 6$ sec, $\dfrac{H_i}{d} = \dfrac{1.5}{3} = 0.5$, $\dfrac{H_i}{gT^2} = \dfrac{1.5}{9.81 \times 6^2} = 0.0043$

From Fig. 8.4, for $\dfrac{H_i}{gT^2} = 0.0043$, $\dfrac{h_o}{H_i} = 0.66$,

$h_o = 0.66 H_i = 1.0$ m

$y_c = d + h_o + \left(\dfrac{1+K_r}{2}\right) H_i = 5.5$ m (as K_r is taken as 1)

$y_t = d + h_o - \left(\dfrac{1+K_r}{2}\right) H_i = 2.5$ m

Compute,

$$\frac{b}{y_c} = \frac{4.5}{5.5} = 0.818, \quad \frac{b}{y_t} = \frac{4.5}{2.5} = 1.80 > 1.0$$

From Fig. 8.9, the reduction factors, r_f and r_m for $b/y_c = 0.818$ are found as,

$$r_f = 0.968 \quad \text{and} \quad r_m = 0.912, \text{ and } \frac{b}{y_t} > 1.0$$

Reduced forces and moments are calculated as

$$F' = r_f F, M' = r_m M$$

Use appropriate r_f and r_m corresponding to crest or trough

At crest,

$$F'_c = 0.968 F_c \quad \text{and} \quad M'_c = 0.912 M_c$$

where F_c and M_c are the forces and moments including that due to hydrostatic component owing to the non-breaking waves. The horizontal wave forces may be evaluated using Fig. 8.5. The value of $F/\gamma d^2$ above $F/\gamma d^2 = 0$ will be the dimensionless force when the crest is at the wall: $F_c/\gamma d^2$, the lower family of curves (below $F/\gamma d^2 = 0$) gives dimensionless forces when the trough is at the wall: $F_t/\gamma d^2$.

with $\dfrac{H_i}{gT^2} = 0.0043$, $\dfrac{H_i}{d} = 0.5$, $\dfrac{F_c}{\gamma d^2} = 0.63$, $\dfrac{F_t}{\gamma d^2} = -0.31$ for $T = 6$ sec

Assuming $\gamma = 10$ kN/m^3,

$$F_c = 56.7 \text{ kN/m or } 5781 \text{ kg/m}$$

$$F_t = -27.9 \text{ kN/m}$$

Total forces including hydrostatic component can be established from Eqs. (8.2) and (8.3).

$$F_{c(\text{total})} = 56.7 + \frac{\gamma d^2}{2} = 101.7 \text{ kN/m}$$

$$F_{t(\text{total})} = -27.9 + \frac{\gamma d^2}{2} = 17.1 \text{ kN/m}$$

The total force acts against the seaward side of the structure, and the resulting net force will be determined by consideration of static loads (e.g. weight of structure), earth loads (e.g. soil pressure behind a seawall) and any other static or dynamic loading which may occur.

Hence, $F'_c = 0.968 \times 101.7 = 98.44$ kN/m

Calculation of moment,

From Fig. 8.6 $\dfrac{M_c}{\gamma d^3} = 0.44$, $\dfrac{M_t}{\gamma d^3} = -0.123$

$$M_c = 118.8 \text{ kN/m}$$
$$M_t = -33.2 \text{ kN/m}$$

The maximum moment at which there is wave action on the leeward side of the structure with $M_c - M_t$

$$M_{net} = M_c - M_t = 152 \text{ kN-m}$$

The combined moment due to both hydrostatic and wave loading is found as,

$$M_{c(total)} = \dfrac{10(3)^3}{6} + 118.8 = 163.8 \text{ kN-m/m}$$

$$M_{t(total)} = \dfrac{10(3)^3}{6} + (-33.2) = 11.8 \text{ kN-m/m}$$

$$M'_c = 0.912 \times 163.8 = 149.4 \text{ kN-m/m}$$

Similarly for trough:

$F'_t = r_f F_t$, $M'_t = r_m M_t$
$r_f = 1.0$ and $r_m = 1.0$ under trough (as $b/y_t > 1$, refer Eqs. (8.10) and (8.12))

$F'_t = F_t$, $M'_t = M_t$ which are calculated earlier respectively as 17.1 kN/m and 11.8 kN-m/m.

Hence, $F'_c = 98.5$, $M'_c = 149.4$ and $F'_t = 17.1$, $M'_t = 11.8$

$$F'_{net} = F'_c - F'_t = 81.4 \text{ kN/m}$$
$$M'_{net} = M'_c - M'_t = 137.6 \text{ kN-m/m}$$

A similar analysis for 10 sec wave gives,

$$F'_{net} = 85.2 \text{ kN/m}$$
$$M'_{net} = 139 \text{ kN-m/m}$$

Problem II

If a smooth faced vertical wall rests on a rubble base of height 2.7 m is considered in a water depth of 3 m for the above problem, find the force and moment on the wall with the wave characteristics considered in the Problem I. Height of rubble foundation $b = 2.7$ m, $H_i = 1.5$ m, $d = 3$ m, $T = 6$ sec or 10 sec. Find the force and moment on the given wall on a rubble foundation.

Solution

We know $y_c = 5.5$ m and $y_t = 2.5$ m
For $T = 6$ sec

$$\frac{b}{y_c} = \frac{2.7}{5.5} = 0.49, \quad \frac{b}{y_t} = \frac{2.7}{2.5} = 1.08 > 1.0$$

From Fig. 8.9 for $T = 6$

$$(1 - r_f) = 0.26; \ (1 - r_m) = 0.52 \text{ and}$$

$$\frac{b}{y_t} > 1.0; \ (1 - r_f) = 0.0; \ (1 - r_m) = 0.0$$

$$F_c'' = 0.26 \times F_c \text{ and } F_t'' = 0.0 F_t$$

F_c and F_t are 101.7 and 17.1 as calculated. Hence $F_c'' = 26.5$ kN/m and $F_t'' = 0$.

Similarly moments at point A,

$$[M_A'']_c = 0.52 \times 163.8 = 85.2 \text{ kN-m/m}$$

$$[M_A'']_t = 0.0(11.8) = 0$$

The overturning moment at point B (refer Fig. 8.10 for details)

$$[M_B'']_c = 85.2 - 2.7(26.5) = 13.7 \text{ kN-m/m}$$

$$[M_B'']_t = 0 \text{ kN-m/m}$$

Similarly for $T = 10$ sec it can be calculated.

Problem III

At vertical wall 4.3 m high is sited in seawater with $d_s = 2.5$ m. The wall is built on a bottom slope of 1:20 (m = 0.05). Reasonable wave periods range from $T = 6$ sec to 10 sec. The breaking wave height for 6 sec wave is 2.8 m and for 10 sec is 3.2 m. Find the maximum pressure, horizontal force, and overtopping moment about the toe of the wall for the given slope.

Solution

For $T = 6$ sec, $d = 2.5$ m

$$\frac{d_s}{L_0} = 0.04448 = \frac{2.5}{56.2}$$

$$\frac{d}{L} = 0.08826, \ L_d = 28.3 \text{ m.}$$

$$D = d_s + (L_d) \times 0.05 = 3.9 \text{ m.}$$

$$\frac{D}{L_0} = 0.0694, \ \frac{D}{L_D} = 0.1134, \ L_D = \frac{3.9}{0.1134} = 34.4 \text{ m.}$$

i.e. $L_D \cong 35$ m

From Eq. (8.15),

$$p_m = 101\gamma \frac{H_b}{L_D} \cdot \frac{d_s}{D}(D + d_s)$$

$$p_m = 331 \text{ kN/m}^2$$

A similar analysis for $T = 10$ sec gives, $p_m = 182$ kN/m^2

The above values can be obtained more rapidly by using Fig. 8.12. To use the figure.

$$\frac{d_s}{gT^2} = \frac{2.5}{9.81 \times 6^2} = 0.0071$$

With $m = 0.05$ read $p_m/\gamma H_b$ corresponding to $d_s/gT^2 = 0.0071$ (Fig. 8.12)

$$\frac{p_m}{\gamma H_b^2} = 12, \ \therefore p_m = 12 \times 10 \times 2.8 = 336 \text{ kN/m}^2$$

Similarly for $T = 10$ sec, $p_m = 176$ kN/m^2

The force can be evaluated as

$$R_m = \frac{p_m H_b}{3} = \frac{331 \times 2.8}{3} = 309 \text{ kN/m}$$

$$M_m = R_m d_s = \frac{p_m d_s H_b}{3} = 772 \text{ kN-m/m}$$

Similarly for $T = 10$ sec, the force and moment are calculated as $R_m = 194$ kN/m, $M_m = 485$ kN-m/m.

Problem IV

A vertical wall 3 m high in a water depth of $d_s = 2.5$ m on a near shore slope of 1:20 ($m = 0.05$), $H_b = 2.8$ m for 6 sec wave, $H_b = 3.2$ m for 10 sec wave. Find the reduced force and overturning moment because of the reduced wall height.

Solution

For $T = 6$ sec, $H_b = 2.8$ m
  From previous problem,
$$R_m = 309 \text{ kN/m}, \quad M_m = 772 \text{ kN-m/m},$$
and for $T = 10$ sec, $H_b = 3.2$
$$R_m = 194 \text{ kN/m}, \quad M_m = 485 \text{ kN-m/m}.$$
For the breaker with $T = 6$ sec, the height of the breaker crest above the bottom is
$$\left(d_s + \frac{H_b}{2}\right) = \left(2.5 + \frac{2.8}{2}\right) = 3.9 \text{ m}$$
The value of b' as defined in Fig. 8.13 is 1.9 m, b' is obtained as (H_b minus the height obtained by subtracting the wall crest elevation from the breaker crest elevation).

i.e. $b' = H_b - \left[\left(d_s + \frac{H_b}{2}\right) - \text{wall elevation}\right] = 2.8 - [3.9 - 3] = 1.9$ m.

$$\frac{b'}{H_b} = \frac{1.9}{2.8} = 0.679, \text{ from Fig. } 8.13, r_m = 0.83$$

$$\therefore R'_m = r_m \, R_m = 0.83(309) = 256 \text{ kN/m}$$

From Fig. 8.14 with $\dfrac{b'}{H_b} = 0.679$

$2a/H_b = 0.57$

Hence $a = \dfrac{0.57(2.8)}{2} = 0.8$ m

Hence,
$$M'_m = R_m[r_m(d_s + a) - a]$$
$$M'_m = 309[0.83(2.5 + 0.8) - 0.8]$$
$$M'_m = 600 \text{ kN-m /m}.$$
For $T = 10$ sec, $r_m = 0.79, a = 0.86$
$$R'_m = 153 \text{ kN/m}, \quad M'_m = 348 \text{ kN-m/m}.$$

272 Coastal Engineering: Theory and Practice

Problem V

A caisson of height 6.5 m from the top of a mound in a water depth of 6 m is exposed to waves with $H_s = 1$ m or $H_{\max} = 1.8$ m and $T_{1/3} = 4$ sec. Calculate the forces on the caisson using the formula of Goda (2010) and Miche-Rundgren (1958).

(a)

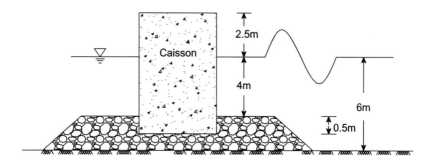

Solution

From dispersion relation → $L = 20.9$ m, $k = 0.301$ m^{-1} at $H_s = 4$ m
Assume non-breaking waves:

(a) Miche-Rundgren (1958)

$$h_o = \frac{\pi \times 1.8^2}{20.9} \coth(0.301 \times 4) = 0.58 \text{ m}$$

$$p_1 = \frac{1+1}{2} \left[\frac{1 \times 1.8}{\cosh(0.301 \times 4)} \right] = 0.99 \text{ t/m}^2$$

$$R_e = \frac{(4 + 1.8 + 0.58)(1 \times 4 + 0.99)}{2} - \frac{1 \times 4^2}{2} = 7.9 \text{ t/m}$$

(b) Goda (2010)

$$\eta* = 0.75(1 + \cos 0)1.8 = 2.7 \text{ m}$$

from dispersion relation → $k = 0.272$ m^{-1} at $d = 6$ m

$$h_c = 2.5 \text{ m} < \eta*$$

assume $h_b = d = 6$ m

$$h_w = 6.5 + (4.5 - 4) = 7 \text{ m}$$

$$\alpha_1 = 0.6 + 0.5 \left(\frac{2 \times 0.272 \times 6}{\sinh 2 \times 0.272 \times 6}\right)^2 = 0.63$$

$$\alpha_2 = \min. \text{ of } \frac{6-4}{3 \times 6}\left(\frac{1.8}{4}\right)^2 = 0.0225 \text{ or } \frac{2 \times 4}{1.8} = 4.4$$

$$\alpha_2 = 0.0225$$

$$\alpha_3 = 1 - \frac{7-2.5}{6}\left(1 - \frac{1}{\cosh(0.272 \times 6)}\right) = 0.53$$

$$p_1 = 0.5(1+1)(0.63 + 0.0225 \times 1) \times 1 \times 1.8 = 1.2 \text{ t/m}^2$$

$$p_2 = \left(1 - \frac{h_c}{\eta*}\right) p_1 = \left(1 - \frac{2.5}{2.7}\right) 1.2 = 0.09 \text{ t/m}^2$$

$$p_3 = 0.53 \times 1.2 = 0.6 \text{ t/m}^2$$

$$R = \frac{1}{2}h_c(p_1 + p_2) + \frac{1}{2}(p_1 + p_3)(h_w - h_c)$$

$$= \frac{1}{2}2.5 \times (1.2 + 0.1) + \frac{1}{2}(1.2 + 0.6)(7 - 2.5) = 5.7 \text{ t/m}$$

Horizontal forces estimated from,

Miche-Rundgren is 7.9 t/m and, Goda is 5.7 t/m

(b)

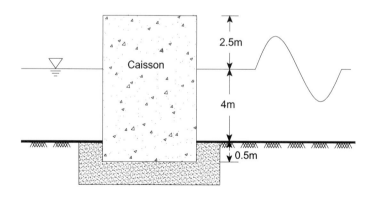

Solution

(i) *Miche-Rundgren*
No change $\rightarrow R_e = 7.9$ t/m

(ii) Goda

$$\eta* = 2.7 \text{ m}$$

$k = 0.301 \text{ m}^{-1}$ at $h_s = 4 \text{ m}$, $h_c = 2.5 \text{ m} < \eta*$, assume $h_b = h_s = 4 \text{ m}$, $h_w = 7 \text{ m}$

$$\alpha_1 = 0.6 + 0.5 \left(\frac{2 \times 0.301 \times 4}{\sinh 2 \times 0.301 \times 4} \right)^2 = 0.7$$

$$\alpha_2 = \text{min. of } \frac{4-4}{3 \times 4} \left(\frac{1.8}{4} \right)^2 = 0 \text{ or } \frac{2 \times 4}{1.8} = 4.4$$

$$\alpha_2 = 0$$

$$\alpha_3 = 1 - \frac{7 - 2.5}{4} \left(1 - \frac{1}{\cosh(0.301 \times 4)} \right) = 0.49$$

$$p_1 = 0.5(1+1)(0.7+0) \times 1 \times 1.8 = 1.3 \text{ t/m}^2$$

$$p_2 = \left(1 - \frac{h_c}{\eta*} \right) p_1 = \left(1 - \frac{2.5}{2.7} \right) 1.3 = 0.09 \text{ t/m}^2$$

$$p_3 = 0.49 \times 1.3 = 0.6 \text{ t/m}^2$$

$$R = \frac{1}{2} 2.5 \times (1.3 + 0.09) + \frac{1}{2}(1.3 + 0.6)(7 - 2.5) = 5.9 \text{ t/m}$$

Horizontal forces estimated from,

Miche-Rundgren is 7.9 t/m.
Goda is 5.9 t/m

i.e., Miche-Rundgren formula estimates about 1.3 times that predicted by the formula of Goda.

Problem VI

Design a rubble mound breakwater in a water depth (d) of 8 m under the 50-year wave climate of 5 m (H_{s50}) and 8 sec (T_m). Consider, the highest high water level (HHWL) is 2.4 m with a storm surge of 0.5 m.

Solution

Total water depth (d_{total}) in front of the breakwater is estimated as,

$$d_{total} = d + \text{HHWL} + \text{Storm surge} = 10.9 \text{ m}$$

Hence, maximum sustainable wave height, $H_{\max} = 0.78 \times d_{total} = 8.5 \text{ m}$.

⇒ Sustainable significant wave height $= H_{\max}/1.8 = 4.72$ m $< H_{s50} = 5$ m

Thus, wave could break on the rubble mound and the structure should be designed for breaking condition having a design wave height (H_{des}) of 4.72 m.

Armour layer design: (Hudson formula)

$$N_s = \frac{H_{des}}{\Delta_a D_a} = (K_D \cot\theta)^{\frac{1}{3}}$$

Let, $\gamma_s = 2.65$ t/m^3; $\gamma_w = 1.025$ t/m^3,

$$\Delta_a = \left(\frac{\gamma_s}{\gamma_w} - 1\right) = 1.585.$$

$K_D = 2$ for breaking condition (CEM, 2006) for a random rubble mound structure with a side slope of $\cot\theta = 2$.

Weight of the armour stone (W),

$$W_{50} = \frac{\gamma_s H_{des}^3}{K_D \Delta_a^3 \cot\theta} = 17.49 \text{ ton}$$

Adopt $W_{50} = 18$ ton with a gradation of (75% to 125%) W_{50}, i.e., the range of armour layer stones is 13.5 ton to 22.5 ton and placed in two layers.

Thickness of armour layer

$$t_{armour} = nK_\Delta D_{50}$$

where, n is the number of layers and the layer coefficient, K_Δ can be obtained from CEM (2006) for different types of armour units and D_{50} is the characteristic dimension of an armour unit.

The characteristic dimension of a single armour unit is defined as the equivalent diameter, defined as,

$$D_{50} = \sqrt[3]{\frac{W_{50}}{\gamma_s}}, \text{ i.e., } D_{50} = 1.89 \text{ m}$$

Let, $K_\Delta = 1$, then, $t_{armour} = 3.787$ m.

Underlayer design

The requirement of underlayer depends on the size of armour stones. Let us assume, one underlayer is required. If more than one underlayer is required, the same design procedure will be repeated for each underlayers.

The weight of first underlayer stones is in the range of $\dfrac{W_{50}}{10}$ to $\dfrac{W_{50}}{15}$.

Adopt, the range of underlayer stones with weights of 1.8 ton to 1.2 ton.

Thickness of underlayer: For any layer made up of rubble stones, there should be minimum two layers. Let us adopt, two layers of underlayer stones. Following the armour layer thickness calculation with a layer coefficient of one,

$$t_{ul} = 1.7 \text{ m}$$

The diameter of underlayer stone is in the range of 0.77 m to 0.88 m.

Core layer

The weight of core layer stones in the range of $\dfrac{1}{10}$ to $\dfrac{1}{15}$ of the layer stone weight placed above it.

Hence, here, the core stone weights are in the range of $\dfrac{W_{50}}{100}$ to $\dfrac{W_{50}}{225}$, i.e., 180 kg to 80 kg.

Filter layer

Following the earlier step, the weights of filter layer stones which distributes the entire rubble stone structure weight into the sea bed is $\dfrac{1}{10}$ to $\dfrac{1}{15}$ of the core layer. Further, if the above design range does not satisfy with reference to the seabed sediment characteristics, additional underlayer(s) need to be introduced to reduce further the filter stone weight. Further, to avoid placing stones of size less than ten times that of a typical seabed sediment mean diameter, the minimum size of stones in the filter layer is defined in terms of linear dimension instead of weight. Hence, adopt the range of filter layer stones of 10 mm to 18 kg. The filter layer thickness is adopted as 300 mm.

Toe layer

The range of stone weights in the toe layer is $\dfrac{W_{50}}{10}$ to $\dfrac{W_{50}}{15}$, i.e., 1.8 ton to 1.2 ton. Minimum width of toe is three layer thickness and the minimum toe height is two layer thickness.

Thus, the width is arrived at 2.637 m and provide 3 m width. Height (thickness) of toe layer is 1.758 m.

Crest elevation

Crest Elevation is calculated as (d_{total}+ Run-up + Free board).
The wave run up is calculated as a function of $\dfrac{H}{gT^2}(= 7.51784 \times 10^{-3})$.
From CEM (2006), $\dfrac{R}{H^o} = 0.8$, $R = 3.776$ m.
Hence, fix, say crest elevation at 15.5 m from sea bed with a free board of 0.5 m.

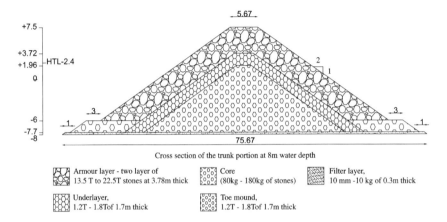

Cross section of the trunk portion at 8m water depth

Armour layer - two layer of 13.5 T to 22.5T stones at 3.78m thick
Core (80kg - 180kg of stones)
Filter layer, 10 mm -10 kg of 0.3m thick
Underlayer, 1.2T - 1.8Tof 1.7m thick
Toe mound, 1.2T - 1.8Tof 1.7m thick

Problem VII

Design a rubble mound breakwater section in a water depth of 18 m for a major port. The highest high tide level is +1.2 m and the storm surge at the location is 1.0 m. The offshore design wave height (H_s) is given as 7.0 m. The seabed is of coarse sand to rocky strata. Adopt the unit weight of rubble stones as 2.55 t/m³, Concrete as 2.4 ton/m³ and water as 1.03 t/m³.

Solution

Given: High tide level (HTL) = 1.2 m, Storm surge = 1 m, Offshore design wave, $H_{offshore} = 7$ m.
Total water depth (d_{total}) in front of the breakwater is estimated as,

$$d_{total} = d + \text{HTL} + \text{Storm surge} = 18 + 1.2 + 1 = 20.2 \text{ m}.$$

Hence, maximum sustainable wave height, $H_{max} = 0.78 \times d_{total} = 15.8$ m.
⇒ Sustainable significant wave height = $H_{max}/1.8 = 8.75$ m > $H_{off} = 7$ m
Thus, the offshore design wave would not break on the structure.

Hence, the structure should be designed for non-breaking wave condition having a design wave height (H_{des}) of 7 m.

Armour design: (Hudson formula)

$$N_s = \frac{H_{des}}{\Delta_a D_a} = (K_D \cot \theta)^{\frac{1}{3}}$$

Let us adopt the concrete armour, Tetrapods with the concrete density, $\gamma_s = 2.4$ t/m^3.

$$\Delta_a = \left(\frac{\gamma_s}{\gamma_w} - 1\right) = 1.33.$$

K_D for tetrapod = 8 for non-breaking condition and for the side slope of $\cot \theta = 1.5$ to 3 (CEM, 2006). Further, adopt the side slope of $\cot \theta = 2$.

Weight of the tetrapod units, $W = \dfrac{\rho_a H_{des}^3}{K_D \Delta_d^3 \cot \theta} = 21.86$ ton.

Hence, adopt 22 ton Tetrapods and the tetrapod units are placed in two layers ($n = 2$) for their stability.

Thickness of the armour layer with a layer coefficient, K_Δ of 1.15 CEM (2006),

$$t_{armour} = n K_\Delta \sqrt[3]{\frac{W}{\gamma_c}} = 4.8 \text{ m}$$

Underlayer

The underlayer can be made with quarry stones to support the armour units. The weight of first under layer stones is in the range of $\dfrac{W}{10}$ to $\dfrac{W}{15}$.

Adopt, the range of under layer stones with weights of 2.2 ton to 1.5 ton. Thickness of under layer for two layers is, $t_{ul} = 1.9$ m.

Now, either the choice of adopting second underlayer or extending the range of core layer may be designed. In the present design, the core layer is designed below the first underlayer since the seabed is made of coarse sand to rocky strata on which higher gradation stones may be placed without appreciable sinkage and hence, the layers will be stable.

Core layer

The weight of core layer stones in the range of $\dfrac{1}{10}$ to $\dfrac{1}{15}$ of the layer stone weight placed above it. Hence, here, the core stone weights are in the range of $\dfrac{W}{100}$ to $\dfrac{W}{225}$, i.e., 220 kg to 100 kg.

Filter layer

Following the earlier step, the weights of filter layer stones which distributes the entire rubble stone structure weight into the sea bed is $\frac{1}{10}$ to $\frac{1}{15}$ of the core layer. Since, the seabed is of stable strata, the similar range of stones is considered, i.e., the range of stone weights in the filter layer is $\frac{W}{1000}$ to $\frac{W}{3375}$. Hence, adopt the range of filter layer stones of 6 kg to 22 kg subject to the minimum size of filter stones as 10 mm. The filter layer thickness is adopted as 300 mm.

Toe layer

The range of stone weights in the toe layer is $\frac{W}{10}$ to $\frac{W}{15}$, i.e., 2.2 ton to 1.5 ton similar to underlayer stones. In a typical design, the under layer is extended to form a toe. Minimum width of toe is three layer thickness and the minimum toe height is two layer thickness. Thus, the width is arrived at 2.85 m, say 3 m and the height (thickness) of toe layer is 1.9 m, say 2 m.

Crest elevation

Crest Elevation is calculated as (d_{total}+ Run-up + Free board).

For wave run up (CEM, 2006), the surf similarity parameter, ξ is calculated.

$$\xi = \frac{\tan \alpha}{\sqrt{S_{op}}}$$

$$S_{op} = \frac{2\pi H}{gT^2} = \frac{2\pi \times 7}{g \times 8^2} = 0.07005$$

$$\xi = \frac{1/2}{\sqrt{0.07005}} = 1.8896$$

From CEM (2006), $\frac{R}{H^o} = 1.0$, $R = 7$ m

Hence, fix, say crest elevation at 28 m from seabed with a free board of 0.5 m. Further, the minimum crest width required is three armour layer unit thickness. Here, provide 7.2 m wide crest. Hence, provide armour layer with tetrapod units and, underlayer, toe berm and core with quarry stones.

Problem VIII

Design a non-overtopping concrete caisson breakwater of rectangular section in a water depth (d) of 8 m. The high tidal level is 2.5 m and the design wave height is 4 m. The horizontal wave force exerted on the caisson is 620 kN/m and the resultant force acts at 6.5 m above the seabed. The maximum run up is given as 4.2 m. The friction (μ) between the concrete and the soil is 0.6.

Solution

The crest elevation is designed such that wave would not overtop.

Crest elevation = d+ HTL + run-up + free board = 15 m with a freeboard of 0.3 m.

Adopt a crest elevation of 15 m from the seabed.

The concrete caisson is designed as mass concrete structure and thus, the structure should be designed to take only compressive loads to minimize the tensile stresses.

$$\text{i.e., } \frac{W}{A} = \frac{M}{Z},$$

where, W is the weight of caisson; A (= width of the caisson, $B\times$ length of the caisson breakwater) is the area of cross-section at the based and Z is the section modulus of the base section with reference to the wave direction.

The moment (M) induced on the breakwater due to wave is (620×6.5) kN-m/m.

Since the breakwater is generally of long structure, the stability of one meter long breakwater is considered. For zero tension requirement, the required width of the caisson (B) is 8.2 m.

$$\frac{W}{A} = \frac{M}{Z} \Rightarrow B = 8.2 \text{ m}$$

In order that the caisson to be stable against sliding and overturning, the following design checks should be made.

Check for sliding

A factor of safety of 1.5 or higher against sliding should be maintained.

$$\frac{\mu W}{F} = 2.85 > 1.5$$

Hence the structure is safe against sliding.

Check for Overturning

To avoid overturning, a minimum factor of safety of 1.5 for resisting moment (M_R) induced by the self-weight of the structure to induced moment (M)

by wave should be maintained.

$$\frac{M_R}{M} > 1.5 \text{ i.e., } \frac{M_R}{M} = \frac{WB/2}{620 \times 6.5} = 3 > 1.5$$

Hence the structure is safe against overturning.

Problem IX

Design a diaphragm wall capable of holding the land and water apart through a combination of its own mass with the passive resistance of the ground forming the seabed immediately in front of it. Consider the following parameters for design: Grade of concrete = M35, Grade of Steel = Fe500, Clear cover = 75 mm, Angle of internal friction = 38°, Dry unit weight of soil = 19.5 kN/ m², Saturated unit weight of soil = 20.5 kN/m², Submerged unit weight of soil = 9.5 kN/m², Surcharge live load = 30 kN/m², Active earth pressure coefficient (k_a) = 0.237, Passive earth pressure coefficient (k_p) = 4.2. Consider the berthing load as 100 kN/m and mooring load as 15 kN/m. The height of the wall needs to be maintained at 6.8 m. Also estimate the tension in tie rod to be placed at 1.5 m from top level of quay wall.

Solution

The dimensions of the wall are assumed as given in the drawing.

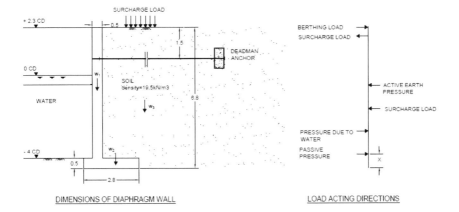

DIMENSIONS OF DIAPHRAGM WALL   LOAD ACTING DIRECTIONS

Calculation of pressures:

1. Active earth pressure = $K_a \times$ unit weight of soil $\times$ height
   = $0.237 \times 19.5 \times 6.8$
   = 31.42 kN/m².
2. Pressure due to surcharge load = 0.237×30 kN/m².
   = 7.11 kN/m²
3. Pressure due to water = $10.03 \times 4 = 40.12$ kN/m².

Calculation of forces:

(1) Active earth pressure = $0.5 \times 6.8 \times 31.42 = 106.83$ kN/m. acting @ 2.27 m from bottom as variation of pressure distribution is triangular
(2) Force due to surcharge load = $0.5 \times 6.8 \times 7.11 = 24.174$ kN/m. @ 2.27 m from bottom
(3) Force due to water = $40.03 \times 0.5 \times 4 = 80.06$ kN/m. @ 1.33+0.5 m from bottom
(4) Berthing force = 100 kN/m. @ 6.8 m from bottom
(5) Mooring force = 15 kN/m. @ 6.8 m from bottom

Calculation of Resultant force:

Both mooring and berthing force would not impart at the same time.
Considering only the Mooring force,
Active resultant force, $R_a = +106.83 + 24.174 - 80.06 + 15$,

$$R_a = 65.94 \text{ kN/m}.$$

Similarly, when berthing force alone is taken into account,
Total resultant force, $R_a = 106.83 + 24.174 - 80.06 - 100$

$$R_a = -49.05 \text{ kN/m}.$$

Design is carried out for the critical case having high magnitude of force. Hence we consider mooring force only case.

Calculation of centroid distance:

To find point at which resultant force acts,

$$\bar{y} = \frac{106.83 \times 2.27 + 24.174 \times 2.27 - 80.06 \times 1.833 + 15 \times 6.8}{65.94}$$

$\bar{y} = 3.83$ m from bottom of the wall.

Calculation of passive force offered:

Height which offers passive resistance to wall $= x$.

$$R_p = 0.5 \times \bar{\gamma}(K_p - K_a) \times x^2$$
$$= 0.5 \times (19.5 - 10) \times (4.2 - 0.237) \times x^2$$
$$= 18.82\ x^2$$

Calculation of tension force in anchor rod:

Anchor rod is placed at 1.5 m from top of the wall. Moment about anchor level,

$$R_a(\bar{y} - 1.5) = R_p((6.8 - 1.5) + 2/3x),\ x = 1.18\ \text{m}$$
$$R_p = 26.27\ \text{kN/m}$$

Tension in tie rod, $T = R_a - R_p$

$$= 65.94 - 26.27 = 39.67\ \text{kN/m}$$
$$\approx 40\ \text{kN/m}$$

Check for overturning:

$M_o = $ overturning moment caused by forces acting on wall
$M_o = $ moment due to active pressure $-$ moment due to passive pressure $-$ moment due to water $+$ moment due to surcharge $-$ tension in tie rod
$\quad = 106.83 \times 2.27 - 16.27 \times 5.92 - 80.06 \times 2 + 24.174 \times 2.27 - 49.67$
$\quad = 87.57\ \text{kNm}$
$M_R = $ Resisting moment due to downward forces
$\quad = W_1 + W_2 + W_3$
$\quad = 0.5 \times 6.3 \times 25 + 0.5 \times 2.8 \times 25 + 6.3 \times 19.5 \times 1.8$
$\quad = 331.13\ \text{kNm}$

$$F = \frac{M_R}{M_o} = 1.8 > 1.5.\ \text{Hence Safe.}$$

Sliding check:

$$\text{Active pressures} = 0.5 \times 0.237 \times 19.5 \times 6.8^2 + 30 \times 0.237 \times 6.8$$
$$= 155.197$$

$$\text{Maximum possible friction force} = \mu W = 0.6 \times 331.13$$
$$= 198.678$$

Passive pressure $= 16.27[(6.8 - 1.5) + (2/3 \times 0.97)]$

$$= 96.31$$

$$\text{Check for sliding} = \frac{\mu W + \text{Passive pressure}}{\text{Active pressure}} = \frac{198.678 + 96.31}{155.197}$$

$$= 1.9 > 1.5. \text{ Hence Safe.}$$

Problem X

Design the diaphragm wall for the critical case under three cases. Case 1: highest high tide level, Case 2: Low tide level and Case 3: high tide level on sea side and empty on land side. Consider the following parameters: grade of concrete = M35, grade of Steel = Fe250, clear cover = 75 mm and the dead loads, water: 10 kN/m³, RCC: 2.5 kN/m³, steel: 7.85 kN/m³. Consider $\gamma_{dry} = 18$ kN/m³, $\gamma_{sat} = 20$ kN/m³, $\gamma_w = 10.25$ kN/m³, $\gamma' = 9.75$ kN/m³ and $\Phi = 32°$.

Case 1: In case of highest high tide level

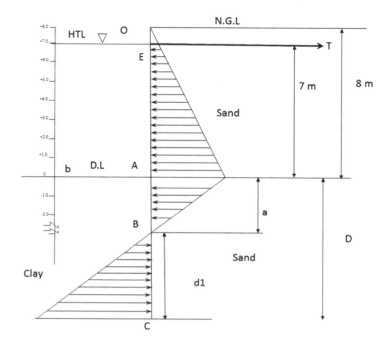

Calculation

Calculate coefficient of active earth pressure

$K_a = \tan^2(45-32/2) = 0.307$
$K_p = 3.25/2 = 1.625$

(As K_p value will not be fully mobilized applying factor of safety 2 to K_p value.)

Calculating the active earth pressure

P_a (at point O) = $(\gamma' \times 0 \times K_a) = 9.75 \times 0 \times 0.307 = 0$ kN/m^2
P_a (at point E) = $(\gamma' \times (H_w - H) \times K_a) = 18 \times 1 \times 0.307 = 5.526$ kN/m^2
P_a(at point A) = $5.526 + 20.95 = 26.47$ kN/m^2
P_a (at point B) = 0 kN/m^2

To find a

$$P_a(@A) = \gamma'(K_p - K_a) \times a$$
$$26.47 = 9.25 \times (1.625 - 0.307) \times a$$
$$a = 2.18 \text{ m}$$

Calculating the passive earth pressure

$$P_p(\text{at point C}) = \gamma'(K_p - K_a) \times d1 = 12.192 d1 \text{ kN/m}^2$$

Calculating R_a and R_p:

$$R_a = (1/2 \times P_a(@E) \times (H_w - H)) + (1/2 \times P_a(@A) \times (H_w))$$
$$+ P_a(@E \times H_w + 1/2 \times P_a(@A \times H_w + 1/2 \times P_a(@A) \times a$$
$$= 143.67 \text{ kN}$$

$$R_p = 1/2 \times P_p(@C) \times d1 = 6.26 d_1^2$$

Summation of all horizontal forces

$$\Sigma FH = 0 = R_a - R_p - T = 0$$

Summation of all moments at the tie rod

$$\Sigma M = 0(@ \text{ Tie Rod})$$
$$= 4.283 d1^3 + 58.98 d1^2 - 720.32 = 0$$

By solving this equation we get, $d1 = 3.15$ m

$$D = d1 + a = 3.15 + 2.18 = 5.33 \approx 5.4 \text{ m}$$

Increase it by 20%–40%

$$D_{\text{actual}} = 5.4 \times 1.4 \text{ (assuming as 40\%)}$$
$$D_{\text{actual}} = 7.56 \text{ m}$$
$$D_{\text{actual}} = 5.4 \times 1.3 \text{ (assuming as 30\%)}$$
$$D_{\text{actual}} = 7.02 \text{ m}$$

Finding T:

$$R_p = 6.425 d_1^2$$
$$= 6.425 \times 3.15^2$$
$$R_p = 63.75 \text{ kN}$$

We know

$$R_a - R_p - T = 0, \; T = 143.67 - 63.75 = 79.91 \text{ kN} \approx 80 \text{ kN}$$

Finding the zero shear point

Equating in such a way that shear force is zero (considering x will be below Tie Rod)

$$2.763 + 5.526x + 1/2 \times x^2 \times 9.75 = 80$$
$$4.875x^2 + 5.526x + (-77.237) = 0$$

By solving we get

$$x = 3.45 \text{ m (below tie rod)}$$

Maximum Bending moment

$$M_{\max} = 80 \times 3.45 - 2.763 \times (1/3 + 3.45) - (5.526 \times 3.45 \times 3.45/2)$$
$$- (1/2 \times 9.75 \times 3.45^2 \times (2/3 \times 3.45))$$
$$M_{\max} = 100 \text{ kNm}$$

Calculating required section modulus

$$Z' = M_{\max}/\text{allowable stress}$$
$$= 100 \times 10^6 / 0.67 \times 250 \text{ (In case of Fe250 grade steel)}$$
$$= 597 \text{ cm}^3$$

Section modulus $Z' = 597$ cm^3

Case 2: In case of low tide level

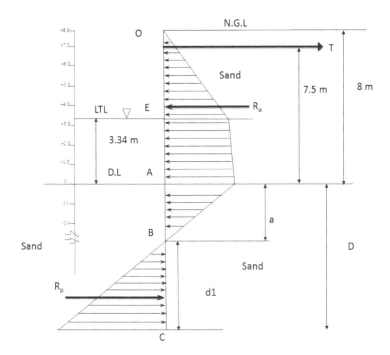

Calculation

Calculating coefficient of active earth pressure

$$K_a = \tan^2(45 - 32/2) = 0.307$$
$$K_p = 3.25/2 = 1.625$$

(As K_p value will not be fully mobilized applying factor of safety 2 to K_p value)

Calculating the active earth pressure

P_a (at point O) $= 0$

P_a (at point E) $= (\gamma' \times (H_w - H) \times K_a) = 25.75$ kN/m^2

P_a (at point A) $= 25.75 + 10 = 35.75$ kN/m^2

P_a (at point B) $= 0$

To find a

P_a (@ A) $= \gamma'(K_p - K_a) \times a$

$35.75 = 9.75 \times (1.625 - 0.307) \times a$

$a = 2.782$ m

Calculating the passive earth pressure

P_p (at point C) $= \gamma'(K_p - K_a) \times d1 = 12.85 d1$ kN/m^2

By solving active and passive pressure we get, $d1 = 4$ m

$D = d1 + a = 4 + 2.782 = 6.782 \approx 6.8$ m

Increase it by 20% – 40%, provide depth 9 m

Section modulus $Z' = 2866.5$ cm^3

Case 3: In case of high tide level on sea side and empty on land side

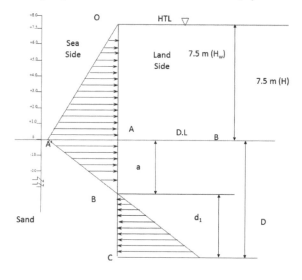

Calculate coefficient of active earth pressure

$K_a = \tan^2(45 - 32/2) = 0.307$

$K_p = 3.25/2 = 1.625$

(As K_p value will not be fully mobilized applying factor of safety 2 to K_p value.)

Calculating the active earth pressure

$$P_a \text{ (at point A)} = (\gamma_w \times H_w) = 10.25 \times 7.5 = 76.875 \text{ kN/m}^2$$

$$P_a \text{ (at point B)} = 0$$

To find a

$$P_a \text{ (@ AA}') = \gamma'(K_p - K_a) \times a$$

$$76.875 = 9.75 \times (1.625 - 0.307) \times a$$

$$a = 5.98 \text{ m}$$

Calculating the passive earth pressure

$$P_p \text{ (at point C)} = \gamma'(K_p - K_a) \times d1 = 12.85 d1 \text{ kN/m}^2$$

By solving active and passive pressure we get

$$d1 = 8.6 \text{ m}$$

$$D = d1 + a = 8.6 + 5.98 \approx 14.5 \text{ m}$$

Increase it by 20% – 40% = provide depth 19 m

But the case we have taken is practically not possible, so provide depth 12 m.

Section modulus Z' = 320301.84 cm³

Based on all the three cases the most critical case is the Case 3.

Problem XI

A cantilever sheet pile wall penetrating a granular soil is shown in the figure below. Here, $L_1 = 2$ m, $L_2 = 3$ m, $\gamma_{dry} = 15.9$ kN/m³, $\gamma_{sal} = 19.33$ kN/m³ and $\phi' = 32°$.

(a) What is the theoretical depth of embedment, D?
(b) For a 30% increase in D, what should be the total length of the sheet piles?
(c) What should be in the minimum section modulus of the sheet piles? Use $\sigma_{all} = 172$ MN/m².

Solution

Part a

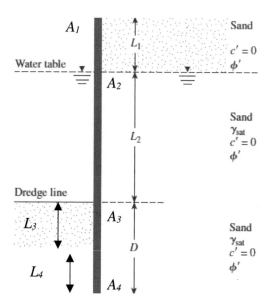

Calculating coefficient of earth pressure

$$K_a = \tan^2(45 - 32/2) = 0.307$$
$$K_p = \tan^2(45 + 32/2) = 3.25$$

Calculation of active pressure

Active pressure above the water table $(\sigma'_1) = \gamma_{dry} L_1 K_a$
$$= (15.9)(2)(0.307) = 9.763 \text{ kN/m}^2$$

Active pressure above the dredge line $(\sigma'_2) = (\gamma_{dry} L_1 + \gamma' L_2) K_a$
$$= [(15.9)(2) + (19.33 - 9.81)(3)(0.307)] = 18.53 \text{ kN/m}^2$$

In order to find the total pressure (P), L_3 need to be calculated,

$$L_3 = \frac{\sigma'_2}{\gamma'(K_p - K_a)} = \frac{18.53}{(19.33 - 9.81)(3.25 - 0.307)} = 0.66 \text{ m}.$$

Total pressure $P = \sigma'_1 L_2 + \frac{1}{2}(\sigma'_2 - \sigma'_1) L_2 + \frac{1}{2}\sigma'_2 L_3$

$$= 9.763 + 29.289 + 13.151 + 6.115 = 58.32 \text{ kN/m}^2$$

Active pressure $(\sigma_3') = (\gamma_{dry}L_1 + \gamma'L_2)K_P + \gamma'L_3(K_P - K_a)$

$= 214.66 \text{ kN/m}^2$

Calculation of centre of pressure by taking moment

$$\bar{Z} = \frac{\sum M}{P} = \frac{1}{58.32}\left[\begin{array}{l} 9.763\left(0.66 + 3 + \dfrac{2}{3}\right) + 29.289\left(0.66 + \dfrac{3}{2}\right) \\ +13.151\left(0.66 + \dfrac{3}{3}\right) + 6.115\left(0.66 \times \dfrac{2}{3}\right) \end{array}\right]$$

$= 2.23 \text{ m}.$

Calculation of L_4

Summation of all horizontal forces

$$\Sigma FH = 0$$

Summation of all moments at the theoretical depth of sheet pile

$$\Sigma M = 0$$

$$L_4^4 + 7.66L_4^3 - 16.65L_4^2 - 151.93L_4 - 230.72 = 0$$

By solving this equation we get, $L_4 = 4.8 \text{ m}$
Thus,

$$D_{theory} = L_3 + L_4 = 0.66 + 4.8 = 5.46 \text{ m}$$

Part b

The total length of the sheet piles is

$$L_1 + L_2 + 1.3(L_3 + L_4) = 2 + 3 + 1.3(5.46) = 12.1 \text{ m}$$

Part c

Finally, we have the following information.

$$z' = \sqrt{\frac{(2)(58.32)}{(3.25 - 0.307)(19.33 - 9.81)}} = 2.04 \text{ m}$$

$$M_{\max} = P(\bar{z} + z') - \left[\frac{1}{2}\gamma' z'^2 (K_P - K_a)\right]\frac{z'}{3}$$

$$= 209.39 \text{ kN-m/m}$$

$$s = \frac{M_{\max}}{\sigma_{all}} = \mathbf{1.217 \times 10^{-3} \text{ m}^3/\text{m} \text{ of wall}}.$$

Problem XII

A steel pipe pile of 61 cm outside diameter with 2.5 cm wall thickness is driven into saturated cohesive soil to a depth of 20 m; the undrained cohesive strength of the soil is 150 kPa (Soil is over consolidated clay). The interior of the pile is filled with M35 mass concrete to add stiffening. The pile is subjected to a lateral load of 268 kN at a height of 2 m above the ground level. Calculate the depth of fixity and the deflection of the pile.
Assume:

$$E_{steel}(E_s) = 208 \times 10^3 \text{ MN/m}^2$$

$$E_{concrete}(E_c) = 20 \times 10^3 \text{ MN/m}^2$$

Solution

To find EI:
    Steel:

$$I_{(steel)} = \frac{\pi(D^4 - d^4)}{64} = \frac{\pi[(0.61)^4 - (0.56)^4]}{64}$$
$$= 1.96 \times 10^{-3} \text{ m}^4$$

$$EI_{(steel)} = (208 \times 10^3)(1.96 \times 10^{-3}) = 407.68 \text{ MN/m}^2$$

Concrete:

$$I_{(concrete)} = \frac{\pi d^4}{64} = \frac{\pi(0.56)^4}{64} = 4.8 \times 10^{-3} \text{ m}^4$$

$$EI_{(concrete)} = (20 \times 10^3)(4.8 \times 10^{-3}) = 96 \text{ MN-m}^2$$

$$EI = EI_{(steel)} + EI_{(concrete)} = 503.68 \text{ MN-m}^2$$

To find the deflection of pile, it is desirable to know the modulus of subgrade reaction (K_h). The modulus of subgrade reaction and stiffness factor are based on the concept of Terzaghi, for more details readers are advised to refer Tomlinson (1994).

$$K_h = \left(\frac{k_1}{1.5}\right)\left(\frac{0.3}{B}\right),$$

B = diameter of the pile, k_1 is Terzaghi's modulus of subgrade reaction as determined from load deflection measurements.

$$k_1 = 27 \text{ MN/m}^3 \text{ (for stiff clay)}$$

$$K_h = \left(\frac{27}{1.5}\right)\left(\frac{0.3}{0.61}\right), \quad K_h = 8.8 \text{ MN/m}^2$$

For soil (clay):

$$\text{Stiffness factor } R = \left[\frac{EI}{K_h B}\right]^{\frac{1}{4}} = \left[\frac{503.68}{(8.8)(0.61)}\right]^{\frac{1}{4}},$$

$$R = 3.11 \text{ m}$$

Then the condition to find the behaviour of pile,

$$L = \text{embedded pile } = 20 \text{ m}$$

$$L \geq 3.5R$$

$$L \geq (3.5)(3.11)$$

$$20 \text{ m} \geq 10.885.$$

Condition satisfied.
Then this is a long pile.

References

Abbot, M. B. and Price, W. A. (1994). *Coastal, Estuarial and Harbour Engineers' Reference Book* (Chapter 29).
Allsop, N. W. H., McKenna, J. E., Vicinanza, D. and Whittaker, T. T. J. (1996). New design methods for wave impact loadings on vertical breakwaters and seawalls, *Proc. 25th Int. Conf. Coast. Eng.*, pp. 2508–2521.
Atkinson, J. (2007). *The Mechanics of Soils and Foundations.* Taylor & Francis.
Bullock, G. N., Crawford, A. R., Hewson, P. J. and Bird, P. A. D. (2000). *Coastal Structures '99*, Vol. 1, pp. 455–463, Balkema, Rotterdam.
Chakrabarti, S. K. (1987). *Hydrodynamics of Offshore Structures*, Springer-Verlag, Berlin.
Coastal Engineering Manual (CEM), (2002). U.S. Army Corps of Engineers, Engineer Manual 1110-2-1100. Washington, D.C.
Coastal Engineering Manual, CEM (2006). U.S. Army Corps of Engineers, Coastal Engineering Research Centre, Vicksburg, Mississippi.
CIRIA (2007). *The Rock Manual: The Use of Rock in Hydraulic Engineering*, CIRIA 683, London.
Code of Practice for Planning and Design of Ports and Harbours (1974) IS 4651: Part 3. UDC 627.2/.3: 624.042.
Coastal Engineering Manual, 2006, Part VI.

Coastal Engineering Research Center (U.S.) (1973). *Shore Protection Manual*, Fort Belvoir, Va.: U.S. Army.
Coastal Engineering Research Center (U.S.). (1984). *Shore Protection Manual, Volume I.*
Dean, R. G. and Dalrymple, R. A. (1991). *Water Wave Mechanics for Engineers and Scientists*, Advanced Series on Ocean Engineering, Vol. 2, World Scientific, Singapore.
Design of sheet pile walls. (1994), CECW-ED Engineer Manual 1110-2-2504, U.S. Army Corps of Engineers.
Goda, Y. (2010). *Random Seas and Design of Maritime Structures*, Advanced Series on Ocean Engineering, Vol. 33, World Scientific.
Goda, Y. (1974). *Coastal Engineering Manual* (2002 in Table VI-5-53, p. VI-5-139)
Hudson, R. Y. (1959). Laboratory investigation of rubble mound breakwaters, *Proc. ASCE*, Vol. 85, No. WW3.
Minikin, R. R., *Winds, Waves and Maritime Structures: Studies in Harbour Making and in the Protection of Coasts*, 2nd ed., Griffin, London.
Port and Harbours — planning and design: Part 2 "Earth Pressures", Bureau of Indian Standards IS 4651.
Punmia, B.C., Ashok Kumar Jain and Arun Kumar Jain (2005). *Soil Mechanics and Foundations*, Lakshmi Publications.
Rundgren, L., Water Wave Forces, Bulletin No. 54, Royal Institute of Technology, Division of Hydraulics, Stockholm, Sweden, 1958.
Saniflou, M. (1928). Treatise on Vertical Breakwaters, *Annals des Ponts et Chaussees*, Paris, France.
Tanimoto, K. and Ojima, R. (1983). Experimental study on wave forces acting on a superstructure of sloping breakwater and on block type composite breakwaters, Tech. Note of Port and Harbour Research Institute No. 450.
Tomlinson, M. J. (1994). *Pile Design and Construction Practice*, Taylor and Francis.
Sundar, V., Vengatesan, V., Anandkumar, G. and Schlenkhoff, A. (1988), Hydrodynamic coefficients for inclined cylinders, *Ocean Engineering*, 25(4–5), 277–294.
Van der Meer, J. W. (2011). Design aspects of breakwaters and sea defences, *5th International Short Conference on Applied Coastal Research.*

Chapter 9

Physical Modeling

9.1 Introduction

Prior to the evolution of computers, physical modeling alone was mainly sought to investigate a wide range of problems related to coastal, harbour and offshore engineering, etc. Even today, physical modeling remain as the only reliable means of understanding the complex phenomena in the field of coastal engineering of which, wave structure interaction, coastal processes, sediment dynamics, near shore characteristics are some of the widespread problems. The random nature of sea state, sediment dynamics, vortex formation, flow through porous media etc. cannot be precisely modeled using numerical techniques, although it has evolved greatly over the years. By means of physical modeling, engineers, researchers, scientists and planners also get a real time picture of the various phenomenon occurring in the vicinity of proposed structures. The various aspects of physical modeling are discussed in detail in this chapter.

9.2 Dimensional Analysis

9.2.1 *General*

The foremost step in physical modeling is to perform a dimensional analysis that considers all the variables associated in defining the problem. The objective of dimensional analysis is to minimize the number of variables and highlight the basis for physical modeling and the physics of similarity. It involves identification of the important independent variables, to arrive at one or more dimensionless parameters which aids to describe problem.

The fundamental representation of all the variables reduces to three basic units namely, mass, length and time. Other physical variables defining a problem related coastal engineering could be height and period of the wave, sediment size, structural characteristics, etc. Table 9.1 presents the details of the basic unit representation for most of the parameters pertaining to coastal engineering.

Table 9.1. Unit representation and scale factor for various parameters.

Variable	Unit	Remarks
GEOMETRY		
Length	L	Any characteristic dimension of the object
Area	L^2	Surface area or projected area on a plane
Volume	L^3	For any portion of the object
KINEMATICS AND DYNAMICS		
Moment of Inertia Area	L^4	
Moment of Inertia Mass	ML^2	Taken about a fixed point
Center of Gravity	L	Measured from a reference point
Time	T	Same reference point (e.g., starting time) is considered as zero time
Acceleration	LT^{-2}	Rate of change of velocity
Velocity	LT^{-1}	Rate of change of displacement
Displacement	L	Position at rest is considered as zero
Angular Acceleration	T^{-2}	Rate of change of angular velocity
Angular Velocity	T^{-1}	Rate of change of angular displacement
Angular Displacement	None	Zero degree is taken as reference
Spring Constant (Linear)	MT^{-2}	Force per unit length of extension
Damping Coefficient	MT^{-1}	Resistance (viscous) against oscillation
Damping Factor	None	Ratio of damping and critical damping coefficient
Natural Period	T	Period at which inertia force = restoring force
Momentum	MLT^{-1}	Mass times linear velocity
Angular Momentum	ML^2T^{-1}	Mass moment of inertia times angular velocity
Torque	ML^2T^{-2}	Tangential force times distance
Work	ML^2T^{-2}	Force applied times distance moved
Power	ML^2T^{-3}	Rate of work
Impulse	MLT^{-1}	Constant force times its short duration of time
Force, Thrust, Resistance	MLT^{-2}	Action of one body on another to change or tend to change the state of motion of the body acted on
STATICS		
Stiffness	ML^3T^{-2}	Modulus of elasticity times the moment of inertia, EI
Stress	$ML^{-1}T^{-2}$	Force on an element per unit area

(*Continued*)

Table 9.1. (*Continued*)

Variable	Unit	Remarks
Moment	ML^2T^{-2}	Applied force times its distance from a fixed point
Shear	MLT^{-2}	Force per unit cross sectional area parallel to the force
Section Modulus	L^3	Area moment of inertia divided by the distance from the neural axis to the extreme fiber
HYDRAULICS		
Kinetic Energy	ML^2T^{-2}	Capacity of a body for doing work due to its configuration
Pressure Energy	ML^2T^{-2}	Energy due to pressure head
Potential Energy	ML^2T^{-2}	Capacity of a body for doing work due to its configuration
Loss to friction	ML^2T^{-2}	Loss of energy or work due to friction
SCOUR		
Grain size of sediments	L	For same prototype material
Free fall velocity	LT^{-1}	Final velocity of a freely falling particle in a medium
WAVE MECHANICS		
Wave Height	L	Consecutive crest to trough distance
Wave Period	T	Time between two successive crests passing a point
Wave Length	L	Distance between two successive crests at a given time
Celerity	LT^{-1}	Velocity of wave (crest, for example)
Particle Velocity	LT^{-1}	Rate of change of movement of a water particle
Particle Acceleration	LT^{-2}	Rate of change of velocity of a water particle
Particle Orbit	L	Path of a water particle (closed or open)
Wave Elevation	L	Form of wave (distance from still waterline)
Wave Pressure	$ML^{-1}T^{-2}$	Force exerted by a water particle per unit area

(*Continued*)

Table 9.1. (Continued)

Variable	Unit	Remarks
STABILITY		
Displacement (Volume)	L^3	Volume of water moved by a submerged object
Righting and Overturning Moment (Hard Volume)	ML^2T^{-2}	Moment about a fixed point of a displaced weight and dead weight, respectively
Natural Period	T	Period of free oscillation in still water due to an initial disturbance
Metacenter	L	Instantaneous center of rotation
Center of Buoyancy	L	Distance of C.G. of displaced volume from a fixed point
MATERIAL PROPERTIES		
Density	ML^{-3}	Mass per unit volume
Modulus of Elasticity	$ML^{-1}T^{-2}$	Ratio of tensile or compressive stress to strain
Modulus of Rigidity	$ML^{-1}T^{-2}$	Ratio of shearing stress to strain

Note:
1. The theory of dimensional analysis is purely algebraic.
2. Problems are expressed by means of dimensionally homogeneous equations.
3. Employ the governing differential equation of a phenomenon or any crude theory to find all variables which may influence the same.

9.2.2 Rayleigh's method

This method determines a parameter for a dependent variable which depends upon maximum of three to four independent variables. If the independent variables exceed four in number, then obtaining a parameter involving the associated dependent variables becomes complex. Let X be a dependent variable which is function of say, X_1, X_2, and X_3 independent variables. Then according to Rayleigh's method,

$$X = f(X_1, X_2, X_3) \tag{9.1}$$

It can be rewritten as

$$X = K\, X_1^a, X_2^b, X_3^c \tag{9.2}$$

where "K" is a non-dimensional constant and a, b, c are arbitrary powers which are obtained by comparing the powers of fundamental dimensions (Dimensional Homogeneity).

Example 1. Find the expression for tidal prism, P in an estuary where H is the average tidal range and A is the average surface area of the basin.

Solution

$$P = K \cdot A^a \cdot H^b \tag{9.3}$$

where, K is a non-dimensional constant
Substitute the dimensions on both sides of Eq. (9.3)

$$M^0 L^3 T^0 = K \cdot (L^2)^a \cdot (L)^b$$

Equating powers of M, L, T on both sides,
Power of L, $3 = 2a + b$
By solving for the above equation, we get $a = 1$ and $b = 1$
Substituting values of a and b in Eq. (9.3)

$$P = K \cdot A^1 \cdot H^1 = A\,H$$

9.2.3 Buckingham's pi theorem

Buckingham's Pi theorem is another popular method of dimensional analysis to obtain the relationship between the variables involved in a given physical problem. Buckingham's Pi theorem states that, *"If there are 'n' variables in a problem and these variables contain 'm' primary dimensions (for example M, L, T) the equation relating all the variables will have (n-m) dimensionless groups"*

The dimensionless groups are represented by π and are called as π groups. According to this theorem, the final expression is given by

$$\pi_1 = f(\pi_2, \pi_3, \ldots, \pi_{n-m}).$$

These π groups formed should be independent of each other and no groups shall be obtained by multiplying powers of other π groups. Compared to the Rayleigh's method of solving simultaneous equations, this method is a lot more advantageous and simpler.

Example. Consider a stationary sphere of diameter D is immersed in a fluid flowing past the sphere in a steady flow where velocity of flow, v; and the fluid properties, i.e., density, r, and viscosity, μ. Therefore, a functional relationship is expected between the drag force F_D and these variables,

$$F_D = \phi(D, v, \rho, \mu) \tag{9.4}$$

An exponential form for this relation is

$$F_D = C D^a v^b \rho^c \mu^d \qquad (9.5)$$

where C is an arbitrary dimensionless constant.

Converting this equation to their dimensional equivalent in an MLT system gives

$$\frac{ML}{T^2} = L^a \left(\frac{L}{T}\right)^b \left(\frac{M}{L^3}\right)^c \left(\frac{M}{LT}\right)^d \qquad (9.6)$$

Equating the exponents of each dimension for dimensional homogeneity, we have

For M: $c + d = 1$
For L: $a + b - 3c - d = 1$
For T: $-b - d = -2$

This gives us 4 unknowns and 3 equations. Writing the equation in terms of one unknown,

$$F_D = C D^{2-d} v^{2-d} \rho^{1-d} \mu^d \qquad (9.7)$$

where C is a constant, rearranging terms,

$$\frac{F_D}{D^2 \rho v^2} = C \left(\frac{vD\rho}{\mu}\right)^{-4} \qquad (9.8)$$

Note that the term within parenthesis is the definition of Reynolds number. Then, the general form of the relationship becomes

$$\overline{F_D} = C\phi(\mathrm{Re}) \qquad (9.9)$$

Where $\overline{F_D}$ is the non-dimensional drag force on the sphere which is represented as a function of Reynolds number (R_e). Note that Eq. (9.4) involved 5 variables in an MLT system so that only $(5 - 3 = 2)$ two non-dimensional quantities are needed (Eq. (9.9)) for a functional relationship. An experiment with a sphere in steady flow produces such a relationship.

Example. Group the variables of a vegetal belt response subjected to both wave and current action. The following variables of vegetal response are of interest for both currents and waves.

Vegetal Response $= f(\rho, g, h, H, T, B_s, D, D_b, BG, f_1, l, SP, E, L, \beta, R_u, V)$

$$(9.10)$$

Where, ρ — mass density of water; g — gravitational acceleration; h — depth of water in the flume; H — wave height; T — Wave period; B_s — Width of the structure; D — diameter of the vegetation; D_b — diameter at the root of the vegetation; BG — width of green belt; f_1 — frequency of first mode of the vegetal stem; l — height of the vegetation; SP — centre to centre spacing between vegetation; E — modulus of plant stiffness; L — Wave length; β — Beach slope; R_u — Run-up; V_f — Flow velocity. The h refers to, h_s — depth of water at the toe of the structure; h_{avg} — average depth of water on the up-stream and downstream of flow in open channel.

The variables seen in Eq. (9.10) are grouped as per Buckingham's Pi theorem. Which, $U_{\max}$ is the velocity V_f and L_0 is the deep water wavelength. The non-dimensional quantities investigated in the study are Darcy's f and Manning's n in steady uniform flow and wave run-up R_u/H, pressures and forces on structures. They are designated as below.

$$f = \left(\frac{8H_f gh}{LV_f^2}\right) \qquad (9.11)$$

$$n = \left(\frac{R^{(2/3)} \times S^{(1/2)}}{V_f}\right) \qquad (9.12)$$

$$F^* = \left(\frac{F_{\max}}{0.5 \times \rho g H^2 B_s}\right) \qquad (9.13)$$

$$P^* = \left(\frac{P_{\max}}{H}\right) \qquad (9.14)$$

9.3 Model Analysis

9.3.1 *General*

Before realizing an engineering project, performing small scale models of the prototype may give you valuable information about the engineering project to be undertaken. This also facilitates understanding the underlying physics behind the problem, although, model studies for a certain problems are forced to be handled with a few limitations. Application of general modeling laws to fluid mechanics problems were started by Reynolds and Froude by conducting a series of experiments to develop a criteria for viscous and inertia effects. Ocean Engineering modeling was extended from that of fluid mechanics taking into account the varying environmental parameters.

The following enlisted are the typical reasons for adopting model studies

- When the known and available analytical methods are insufficient to solve a given problem
- To determine the empirical co-efficient for known equations
- To validate or substantiate the results of a newly introduced analytical tool
- To investigate failure pattern and mechanism.

The terminologies often used in model analysis and their definitions are given below.

Prototype is defined as the actual system or a field problem being investigated.

Model is defined as the substitute system that is proportional to the prototype by a scale ratio.

Homologous is defined as the corresponding conditions in the model and its prototype.

9.3.2 Complete similarity

The priority of most of the model studies is to achieve a simple numerical value such as the destructive or ultimate load. These numerical values are dependent on the variables involved. In order to get accurate results through dimensional analysis the variables are grouped to form a complete set of dimensionless products such as $\pi_1, \pi_2, \ldots, \pi_n$. If the independent variables $\pi_1, \pi_2, \ldots, \pi_n$ possess the same value for the model and the prototype then the model is said to be completely similar. Complete similarity is nearly impossible to achieve in our model studies. It is to be noted that the forces such as surface tension, surface roughness etc., may influence the model but not the prototype and the independent dimensionless variables which are believed to have secondary influences are to be neglected. Scale effects (discussed later in this chapter) of the same magnitude occurs inevitably in all models. To avoid scale effects larger models are to be employed.

9.3.3 Applications of model analysis

Most of the marine structures such as breakwaters, jetties, groins, training walls, seawalls, piers, intake and outfall structures, etc. can be tested, for their interaction with waves as well as their effect on the coast. The behaviour and the stability of coastal structures could be verified in the laboratory by applying a suitable model scale, the results of which could be extended in the field. The extent of sedimentation in a harbour approach channel, influences by wave and tides, behaviour of estuaries, bottom stability of breakwaters and other gravity structures, behaviour of shoreline can be investigated through physical model studies.

9.4 Principles of Similitude

9.4.1 *General*

The behaviour of a model when subjected to external forces holds a definitive relation with that of a prototype. The model should be designed and tested in accordance with certain principles called the principles of similarity (or) principles of similitude. The principles not only govern the design, construction and testing; these principle are applicable to the extrapolate model results in order to predict prototype response.

9.4.2 *Similitude in hydrodynamic problems*

The principle of similitude adopted for hydrodynamic problems need to be treated separately from the conventional similitude. The three major types of similitude viz. geometric similitude, kinematic similitude and dynamic similitude and they are explained in detail as follows. A convention of using "p" as a subscript for prototype and "m" as a subscript for model is followed.

9.4.3 *Types of similitude*

Geometric similitude

Geometric similarity signifies that the dimensional similarity from the prototype to the model i.e. similarity of the physical form. When the horizontal and vertical scales are same, the model is said to be undistorted and when the horizontal and vertical scales not equal the model is said to be a distorted model. Definitive objectives can be achieved by departing from geometric similarity. The major problem areas are confining boundaries of laboratory testing facilities, surface roughness of model, scaling thickness, depth of water, etc. For example, in case of modeling a stretch offshore structure deployed at higher depths, a distorted model is preferred for the study because the horizontal scaling dimension becomes extremely insignificant.

Kinematic similitude

Kinematic similarity is the similarity of motions which implies both geometric similarity and similarity of time intervals, in other words, it requires that the model and prototype have the same length scale ratio and the same time scale ratio; thus the velocity scale ratio will be the same for both.

Example. Estimate the scaling criterion necessary for kinematically similar wave motion for gravity waves whose length is given by

$$L = \frac{gT^2}{2\pi} \tanh \frac{2\pi d}{L}$$

Solution. $2\pi d/L$ is dimensionless, this ratio for model and prototype should be an invariant for geometrically similar model. If $2\pi d/L$ is the same, the tangent of the hyperbola will also be the same therefore scaling relation between L and T is arrived from prototype to model ratio of wave length. Thus for a kinematic similarity of wave motion,

$$\frac{L_p}{L_m} = \frac{\left[\frac{gT^2}{2\pi} \tanh(2\pi d/L)\right]_p}{\left[\frac{gT^2}{2\pi} \tanh(2\pi d/L)\right]_m}$$

Since "g" is constant

$$\frac{L_p}{L_m} = \frac{T_p^2}{T_m^2}$$

$$\lambda_L = (\lambda_T)^2$$

Therefore, the kinematic similarity of wave lengths can be achieved by making **Dynamic Similitude** $\lambda_T = \sqrt{\lambda_L}$.

Dynamic similarity exists when the model and the prototype have the same length scale ratio, time scale ratio, and the force scale (mass scale) ratio. Thus, the ratio of any two forces in one system must be the same as the ratio of the corresponding forces in the other system. Kinematic similarity can be achieved without considering any properties of the fluid, which is unlikely for dynamic similarity.

$$\frac{F_m}{F_P} = \frac{M_m a_m}{M_p a_p} = \frac{\rho_m L_m^3}{\rho_p L_p^3} \times \frac{\lambda_L}{\lambda_T^2} = \lambda_\rho \lambda_L^2 \left(\frac{\lambda_L}{\lambda_T}\right)^2 = \lambda_\rho \lambda_L^2 \lambda_u^2$$

Figure 9.1 depicts a model of OWC showing the similitude between model and prototype scaled down using Froude's scaling law.

9. Physical Modeling 305

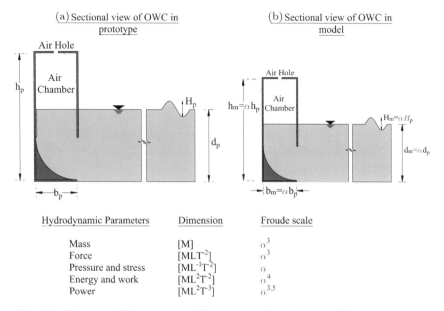

Hydrodynamic Parameters	Dimension	Froude scale
Mass	[M]	α^3
Force	[MLT^{-2}]	α^3
Pressure and stress	[ML^{-1}T^{-2}]	α
Energy and work	[ML^{2}T^{-2}]	α^4
Power	[ML^{2}T^{-3}]	$\alpha^{3.5}$

Fig. 9.1 Geometric, kinematic and dynamic similitude achieved by a model oscillating water column, OWC.

9.5 Scale Effects

Model scales are often derived from dimensional analysis or governing equations. Complete similarity between model and prototype impossible to achieve, since certain quantities such as gravity, cannot be scaled down. The parameters such as gravity, fluid viscosity and density are the same in model as well as prototype. It is impossible to generate both gravity driven and fluid viscous related phenomena concurrently. Waves and currents are gravity-driven forces and in model studies it is preferred to simulate gravity properly. This means that viscosity effects will not be properly reproduced. The effects of such non-similarity are called *scale effect*.

9.6 Model Laws

The laws based on which the models are designed for dynamic similarity are known as model laws or laws of similarity. The model laws are classified based on their application as follows

- Reynold's model law
- Froude's model law

- Euler's model law
- Weber's model law
- Mach's model law

Reynolds model law

Reynolds number is the ratio of inertia force and viscous force. If the viscous forces acting on the system are more predominant than the other forces, the models are designed for dynamic similarity based on Reynolds law, which states that the Reynolds number for the model must be equal to the Reynolds number for the prototype. Models based on Reynolds number includes problems based on pipe flow, resistance experienced by submarines, airplanes, fully immersed bodies etc.

$$R_e = \rho V L / \mu$$

where ρ is the density, V is the velocity, L is the characteristic length, and μ is the viscosity.

Froude's model law

Froude number represents the ratio of inertial forces to gravitational forces. Froude model law is applicable when the gravity force is only predominant force which controls the flow in addition to the force of inertia. Froude number for the model must be equal to the Froude number for the prototype. Froude model law is applicable for free surface flows such as flow over spillways, weirs, sluices, channels etc., flow of jet from an orifice or nozzle and flow of different densities of fluid over one another.

$$F_r = V/(gL)^{0.5}$$

where V is the average velocity, L is the characteristic length associated with the depth, and g is the gravitational acceleration.

Euler's model law

Euler's number represents the ratio of square roots of inertia force to pressure force. This law is applicable when the pressure forces are alone predominant in addition to the inertia force. Dynamic similarity between the model and prototype is achieved when the Euler number for model and prototype are equal. Euler's model law can be applied to fluid flow problems in a closed pipe where turbulence is fully developed so that viscous forces are negligible and gravity force and surface tension force is absent.

$$Eu = \frac{V}{\sqrt{(p/\rho)}}$$

where V is the average velocity, p is the pressure intensity and ρ is the density of the fluid.

Weber's model law

The Weber number denotes the ratio of the inertial forces to surface tension forces. The Weber number becomes an important parameter when dealing with applications involve two fluid interfaces such as the flow of thin films of liquid and bubble formation. Weber model law is the law in which models are based on Weber's number and dynamic similarity between model and prototype can be achieved when the Weber number for both model and prototype are equal. This law is applicable for capillary rise and fall problems.

$$We = \rho V^2 L/\sigma$$

Where σ denotes surface tension

Mach's model law

Mach's number is defined as the square root of the ratio of the inertia force of a flowing fluid to the elastic force. Mach model law is the law in which models are designed on Mach number, which is the ratio of the square root of inertia force to elastic force of a fluid. Hence where the forces due to elastic compression are prominent in addition to inertia force, the dynamic similarity between the model and its prototype is obtained by equating the Mach number of the model and its prototype.

$$Ma = V/c$$

where c is the speed of sound (343 m/s at 20°C) and V is the fluid velocity.

9.7 Case Studies

Physical model studies are widely used in the field of coastal engineering to assess the stability and performance of structures. Typically to initiate a model study, the scaling law to be followed is decided first. Figure 9.2 projects the prototype and model scale down version of a flaring shaped seawall [Kamikubo et al. 2000] constructed at Kurahashi Island, Hiroshima Prefecture at Japan and Fig. 9.3 shows the prototype and model scale down version of curved seawall located at Galveston, Texas, USA [Coastal Engineering Manual, 2006]. Photo 9.1 depicts the fabricated scaled down model of seawall.

Fig. 9.2 Flaring shaped seawall.

Fig. 9.3 Galveston seawall.

The hydrodynamic scaling parameters are water depth, wave height and wave period for a two-dimensional study. For a given scale factor λ, the hydrodynamics parameters are scaled down as follows,

Water depth$_m$ = water depth$_p/\lambda$

$5/10 = 0.5$ m

Wave height$_m$ = wave height$_p/\lambda$

$3/10 = 0.3$ m

Wave period$_m$ = wave period$_p/\sqrt{\lambda}$

$6/\sqrt{10} = 1.89$ sec to $9/\sqrt{10} = 2.84$ sec

9. Physical Modeling   309

Photo 9.1   Image of scaled down curved sea wall.

Now, based on the values arrived a physical model investigation can be performed.

Durgarajapatnam port — Breakwater stability

In order to test the stability of proposed design of breakwaters at the newly anticipated harbour, physical model test were carried out at the 2 m shallow flume facility at the Department of Ocean Engineering, IIT Madras. A model scale of 1 in 24.66 was adopted for the undistorted two-dimensional stability study of the trunk section. Figure 9.4 shows the prototype breakwater cross section and Fig. 9.5 shows the model breakwater cross-section. Table 9.2 gives the details of scaled down unit weights of each layer comprising the breakwater. Table 9.3 gives the details of scaled down wave characteristics for various loading conditions. Photo 9.2 depicts the model section positioned inside the flume.

Stability tests are performed by subjecting the model to random waves of 1000 numbers the time series of which corresponds to P-M spectrum and the results of these tests are given in Table 9.4 as prescribed by the Coastal Engineering Manual.

310 Coastal Engineering: Theory and Practice

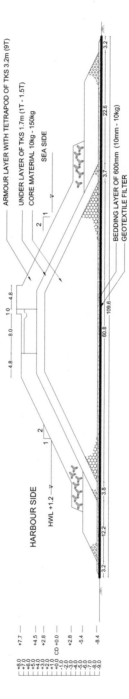

Fig. 9.4 Prototype cross-section of the breakwater.

9. Physical Modeling

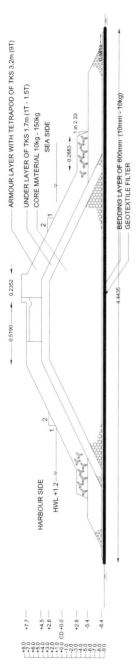

Fig. 9.5 Model cross-section of the breakwater.

Table 9.2. Prototype and model details of the breakwater section.

Layer	Prototype (kg)	Model (gram)
Primary armour	9000	600
Secondary armour	1000 to 1500	67 to 100
Core	10 to 150	0.67 to 10
Filter	1 to 1.5	0.67to 0.1

Table 9.3. Loading conditions for the breakwater at −9m CD.

	Prototype		Model	
Loading	Wave height (m)	Wave period (sec)	Wave height (m)	Wave Period (sec)
Normal	5	9	0.202	1.81
Design	5.4	9	0.218	1.81
Shake down	3.7	9	0.150	1.81
Over loading 1	6.8	9	0.275	1.81
Over loading 2	6.8	9	0.275	1.81

Photo 9.2 View of the model inside the flume.

As a result of the study conducted, no significant damage to the structure was observed and hence it is safe to erect the proposed breakwater section in the field.

Table 9.4. Stability assessment of breakwaters.

Wave period, T		Wave height, H		Water depth		Loading	Movement of armour blocks			Remarks
Prototype (sec)	Model (sec)	Prototype (m)	Model (m)	Prototype (m)	Model (m)		Category	No. of units	% of damage	
9	1.81	5	0.202	10.2	0.41	Normal	P Q R S	1 0 0 1	0.125	One armour unit was rocking
9	1.81	5.4	0.218	11.2	0.45	Design	P Q R S	1 0 0 0	0.125	One armour unit was rocking; Splashing of water drops towards harbour side
9	1.81	3.7	0.150	10.2	0.41	Shakedown	P Q R S	1 0 0 0	0.125	One armour unit was rocking
9	1.81	6.8	0.275	11.2	0.45	Overload 1	P Q R S	4 0 0 0	0.50	Four armour unit were rocking; overtopping was measured by traybox (725 ml)
9	1.81	6.8	0.275	10.2	0.41	Overload 2	P Q R S	1 0 0 0	0.125	One armour unit was rocking

References

Coastal Engineering Manual, CEM (2006). U.S. Army Corps of Engineers, Coastal Engineering Research Centre, Vicksburg, Mississippi.

Kamikubo, Y., Murakami, K., Irie, I. and Hamasaki, Y. (2000). Study on Practical Application of a Non-Wave Overtopping Type Seawall, *Proc. 27th Int. Conf. Coastal Engineering*, Vol. 126, pp. 2215–2228.

Chapter 10

Numerical Modelling

10.1 Introduction

Modelling in general refers to the process of finding solutions for physical problems by suitable methods of approximating the variables involved. It may be of two types: Physical and theoretical modelling. Physical modelling involves laboratory testing of the models or their prototypes under the desired physical conditions to obtain information useful in obtaining any empirical relations, leading to a tangible solution for the problem. Theoretical modelling process consists of forming/defining a mathematical model in a differential or algebraic form, describing the laws of the physical problem. This mathematical model has to be solved for a solution which in most cases is not possible analytically, and hence requiring numerical approximation. The first step in the development of a numerical model is the approximation to the governing equations prescribed by a mathematical model. A systematic calibration of the numerical model is required by tuning its parameters and further, validation against existing data and analytical results provide the confidence to the capability of the model. A prerequisite is the error analysis of the simulated results. The final step is the actual execution of the numerical model to obtain solutions which are analysed and interpreted in the form of graphs, tables, or entailed qualitative forms. Figure 10.1 depicts flow of the steps in obtaining the numerical solution to the given mathematical representation of the problem.

10.2 Need for Numerical Models

Theoretical calculations for obtaining solutions are applicable only in a few practical cases. In most cases, the existing analytical solutions may be too complex to arrive at. It is quite rare that these problems be solved in closed form. Even when closed-form solutions do exist, their behaviour may still be difficult to understand. In order to obtain a better insight to the problem, solutions are generally approximated numerically using discretization methods.

316   Coastal Engineering: Theory and Practice

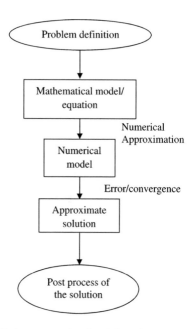

Fig. 10.1   Various steps in obtaining the numerical solution.

Experimental investigations of any physical problem, on the other hand, are expensive and are often impracticable to conduct on a full scale. Even though the physical restraints are simplified with the usage of a scaled model in the experimental investigation, there is always a chance of measurement errors as the extrapolation of the results may be erroneous due to incorrect approximation of the associated physics and thus similarity may not be maintained in the results. Numerical solutions are available for virtually all problems, although available only at discrete points rather than a continuous solution in the case of analytical modelling. Once a model is set up, numerical modelling allows us to simulate different trials with lesser time and cost than in the case of experiments which involve a separate set up for each case. Also, it provides complete information whereas in experiments it is difficult to measure all the parameters. These models can handle any degree of complexity, namely hazardous conditions, but the solution obtained is only as good as the input given.

Since the numerical models are purely based on the approximation of some mathematical equations, the solution may not be in accordance with reality. The models cannot guarantee a correct description of the concerned physical processes involved in the problem if not properly calibrated. Also, there may be a case of existence of multiple solutions. Hence, numerical solution is not a substitute for experiments, without which the numerical models cannot be validated, but can dispense with its shortcomings. The computer simulation of a numerical model provides a more refined dataset which better represents the reality.

10.3 Mathematical Description of Flows: Governing Equations

The concepts of describing the fluid dynamics are built on the fundamental conservation laws namely, the equations of conservation of mass, conservation of energy and conservation of momentum. It is from these concepts that a practical situation is assigned a corresponding governing equation, which is further approximated using numerical methods to a numerical model. These governing equations are listed as below.

(i) *Continuity equation* based on the principle of mass conservation.
(ii) *Energy equation* in adherence with the principle of conservation of energy.
(iii) *Momentum equation* following Newton's second law, $F = ma$.

10.3.1 *Continuity equation*

In fluid dynamics, the principle of conservation of mass, leads to the continuity equation as below.

$$\frac{\partial pu}{\partial x} + \frac{\partial pv}{\partial y} + \frac{\partial pw}{\partial z} = -\frac{\partial \rho}{\partial t} \tag{10.1}$$

$$\frac{\partial \rho}{\partial t} + \nabla \cdot (\rho u) = 0 \tag{10.2}$$

where, (x, y, z) are spatial and t is the temporal independent coordinates; ρ is the fluid density; and, $u = u\vec{i} + v\vec{j} + w\vec{k}$ is the velocity vector.

The time derivative can be understood as the addition or loss of mass over time within the referenced domain. Fluid assumed to be incompressible will imply that $\nabla \cdot u = 0$.

10.3.2 *Momentum equation*

The momentum equations derived from Newton's second law of motion, $F = ma$ are called as Navier–Stokes equations (named after Claude-Louis

Navier and George Gabriel Stokes). These describe the motion of fluid particles.

The forces constitute of body forces, acting on the volumetric mass of the element and surface forces acting on the boundary of the fluid element, due to pressure distribution and shear stress distribution on the surface. For incompressible Newtonian fluids under laminar flow, the momentum equations are written as,

$$\rho\left(\frac{\partial}{\partial t}u + u \cdot \nabla u\right) = -\nabla p + \mu \nabla^2 u + \rho g \quad (10.3)$$

The derivation of analytic solution for Eqs. (10.2) and (10.3) is feasible only for simple domain geometries and for the linear force components. In most of the real field conditions, a direct analytic solution is not possible due to complex geometries and flow conditions. Hence, approximate solutions through numerical discretization is always sought to analyse for flow characteristics.

10.4 Discretization

Numerical approximation provide solutions only at discrete points in the domain rather than continuous variation of the variable in the domain as in the case of analytical solutions. These discrete points are named as *grid points* or *nodes*. The grids may be structured or unstructured; structured grid refers to that grid which follows certain geometrical regularity, whereas the unstructured grid is the one in which the grid points are placed randomly. Discretization methods can be broadly classified as Explicit and Implicit methods. Briefly, an explicit method obtains the successive values at say, t_{n+1} parametrically in terms of given or previously computed quantities at time t_n, the advantage of it being simplified solution algorithms, as it involves direct computation without solving any system of equations. An implicit method is the one which calculates the values at t_{n+1} by using unknown values at t_{n+1} and has to be solved simultaneously at all the grid points as a system of algebraic equations. The latter usually involves inversion of a matrix equation either directly or iteratively at each time step.

The discretization is an art bounded by the consistency and convergence requirements. Further, under the control of conservation laws, the stability of the prescribed numerical scheme depends on the discretization. A discretization scheme is consistent if the difference between partial differential equations (PDEs) and its finite increment minimizes as the grid is refined. And, it is convergence if the discretized equations of the given differential equation solves towards the exact solution as the grid spacing reduces. However, the numerical scheme for utilizing the discretization should not intensify the errors during the process towards the solution and it dictates the stability, i.e., if the ratio of the error in the ith grid point at $(n+1)$th step, ε_i^{n+1} with at the nth step, ε_i^n should be less than or equal to one. By enforcing conservation laws, the numerical scheme is usually been framed from the PDEs which in turn, have been built on the conservation laws. However, the errors due to coarse grids in general are due to non-conservation but these are difficult to quantify.

10.4.1 Discretization techniques

10.4.1.1 Finite difference method (FDM)

Finite difference method involves approximation of the governing equations in differential form with difference equations obtained by replacement of the derivatives with finite, algebraic differences quotients. The domain is first divided into grid, followed by the approximations of the continuous functions at finite nodal values, i.e., one algebraic equation per degree of freedom at a given grid node. Taylor series approximation is the most sought after method of discretization, especially for structured grids. Further the linear system of simultaneous algebraic equations is solved to obtain the values at the grid nodes only.

10.4.1.2 Finite volume method (FVM)

Finite volume method comes into play when an integral representation of the governing equations can be formed. The problem domain is discretized into a finite number of control volume connected together and the conservation laws are enforced to each control volume. Here, the Gauss theorems are employed to convert the volume integrals. The node of the grid is generally located at the centroid of each of the control volume. Many computational fluid dynamics packages make use of this method as it has the added advantage of easy adaptability for unstructured meshes, even for complex geometries.

10.4.1.3 Finite element method (FEM)

The finite element method formulates the solution to a problem by subdividing the domain into a finite number of elements. The governing equations along with the boundary conditions are transformed to prescribe within each element and then the overall solution is obtained after assembling the elements with the continuity requirements between the elements. The finite element system of equations within each element is derived either using weighted residual technique or variational formulation. Due to its flexibility of this approach, this method has been widely adopted to prescribe the flow.

The following sections explain the various numerical modelling schemes being carried out in coastal applications.

10.5 Numerical Wave Modelling

A numerical wave model is arrived at by a combination of the mathematical form which describes the physical wave problem and numerical approximation of the mathematical equations. It differs from theoretical modelling in terms of the method instrumental in reaching the solution of the governing equations. The foundation of numerical wave modelling is laid by the existing wave equations resulting from theoretical studies. Many a times, the same condition of wave propagation may be represented by means of more than one equation, the choice of which depending on the levels of approximation desired. Likewise, a wave model may be applicable to different cases of wave motion provided, the fundamental assumptions hold good.

There are two approaches in defining numerical wave models: the first approach is the resultant of the vertical integration of the time-dependent mass and momentum balance equations that leads to phase-resolving models and the second one is the phase-averaged approach by including the required physical processes into the energy or action balance equation. Appropriate numerical techniques need to be employed to solve equations from either approach.

The phase-resolving models can be applied over relatively small regions of the order of tens of kilometre and the temporal resolution dictates the need for 10 ~ 100 time steps for each wave period. In these models the sea surface elevation is resolved with a grid laid on the free surface. However, the phase averaged models can be used to cover much larger regions and only, the statistics of the sea surface elevation is computed, i.e., in the frequency domain.

10.5.1 *Wave spectral models*

Best suited to model large scale wave propagation, the wave energy spectral model is formulated based on the assumption that an irregular sea state is described as the summation of infinite number of linear monochromatic waves represented by a unique wave height as a function of wave frequency and the propagation direction. For an individual wave train, the rate of change of wave energy (or action) flux is balanced by the wave energy transfer among different wave components in different directions (i.e., wave refraction) and different frequencies (i.e., nonlinear wave interaction) as well as energy input and dissipation. This wave energy spectral model, when linked to an atmospheric model, may be adopted to forecast global ocean wave climate. One of the earlier attempts on wave spectral model is WAM (WAve prediction Model) (Hasselmann et al., 1988). This model balances the evolution of the wave spectrum with the sum of the local wind input, wave dissipation and nonlinear wave-wave interaction with the consideration of incoming swell that are from non-local sources. The source terms describing the wind input, nonlinear transfer of energy, dissipation due to wave breaking, bottom dissipation, and refraction for finite-depth water are prescribed explicitly. The model is formulated in spherical longitudinal and latitude coordinates for an arbitrary region.

The non-inclusion of the diffraction and application of linear wave theory for describing the wave characteristics are the limitations in the application of this approach in the nearshore. This warrants the domain of interest be a few wave lengths away from the barrier and that the corrections for the non-linearities be accounted for. With the above stated background, SWAN — has been developed [Booij et al., 1999; Ris et al., 1999]. SWAN-Simulating Waves Nearshore (Ris et al., 1999) is a wave spectral model which incorporates the effect of currents into the existing wave model to simulate the wave-current interaction. Since the wave phase data is omitted in these types of formulations, they are enabled to generate meshes much larger than a wavelength, thus made suitable for modelling large domains, especially deep oceans. SWAN can be readily nested in WAM. Since the wave phase information is left out of this model, the wave action pertinent to wave diffraction cannot be formulated and thus is left to the scope of other models.

10.5.2 Test case: Wave propagation over constant depth bathymetry

The wave generation over a restrcited water body of $20° \times 20°$ was set up for executing the WAM. A water depth of 250 m was assumed over a region. The grid resolution was fixed as $1/12° \times 1/12°$. The model was executed for different combinations of wind and current fields as mentioned below.

(1) Constant wind blowing over the entire region in the absence of current field
(2) Constant wind in addition to in-line current field
(3) Constant wind over opposing current field

A 10 m/s constant northerly wind was assumed to blow over the region. An initial wave was set up with the same wind condition and the simulation carried out for forty-eight hours until the steady state was attained. In the second case, with the above wind field, a constant in-line current of 5 m/s was assumed to be present. In the last case, a constant opposing current field of 5 m/s was assumed. The current direction in the second condition was the same as the wind direction while, in the last condition, it was 180° out-of-phase with the wind direction.

The simulated wave field was analysed for estimated spectral parameters, significant wave height, H_s and peak wave period, T_p. The variation of the said two wave characteristics which approaches asymptotic values are projected in Fig. 10.2. A comparison of wave characteristics derived from WAM with the wave spectral model and SMB prediction curves [Sverdrup and Munk (1947), Bretschneider (1952, 1958), US Army coastal Engineering Research Centre (1977)] is presented in Table 10.1 for the constant wind condition. Table 10.2 presents the variation in the wave parameters in the presence of in-line and opposing current field are reported in Table 10.2, whereas, the spectra for the three cases are shown in Fig. 10.3. It can be seen that the in-line current field reduces the wave height and shifts the peak frequency towards higher harmonics. The opposing current field however resulted in the steepening of the waves by focusing on the narrow band of frequencies. The frequency components were shifted towards lower harmonics.

Table 10.1. Variation of simulated wave estimates under the action of constant wind and current fields.

S. No.	External forcing	WAM	Wave spectral model	SMB
1.	Significant Wave height	2.14	2.14	2.14
2.	Peak wave period	7.44	7.4	7.69

Table 10.2. Comparison of simulated and analytically derived wave estimates in a constant northerly wind of 10 m/s.

S. No.	External forcing	H_s (m)	T_p (s)
1.	Wind	2.14	7.44
2.	Wind + Inline current	2.07	5.09
3.	Wind + Opposing current	2.10	13.19

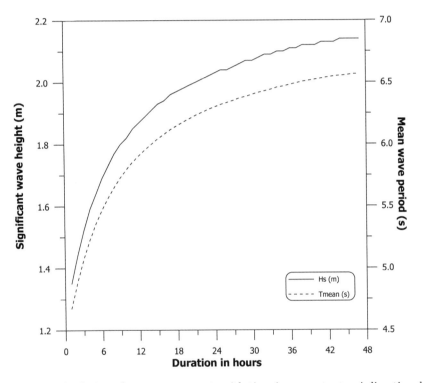

Fig. 10.2 Evolution of wave components with time in a constant uni-directional wind field blowing over the entire region.

10.5.3 Test case: Wave prediction over Bay of Bengal during a cyclone

Given the constraint of the numerical modelling capabilities, the wave prediction accuracy depends on the resolution in the bathymetry and wind characteristics. The cyclonic wave climate off Chennai coastal region,

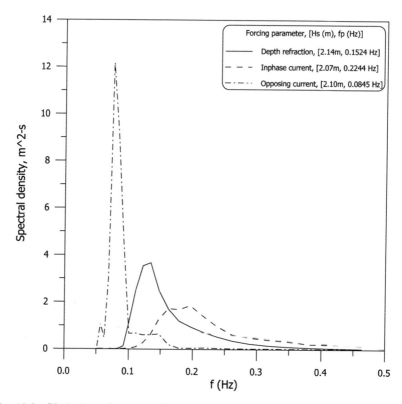

Fig. 10.3 Variation of generated wave spectra under different wind and current fields.

0°–25°N and 75°E–95°E along the south-east coast of India has been simulated with grid resolution of 0.1° × 0.1° using WAM as shown in Fig. 10.4. The ENCEP wind data with a grid resolution of 0.5° × 0.5° from NOAA during the passage of Thane cyclone south of Chennai coast during 12–31 December 2011 was used for simulating wave characteristics. A typical wind vector over the domain at 0600 hours on 29 December 2011 is shown in Fig. 10.5. Fortunately, at the time of the passage of the cyclone the wave characteristics were captured that facilitated a comparison of significant wave height (H_s), mean wave period (T_m) and mean wave direction (θ_m) with that obtained through the directional spectra that were deduced from buoy and directional tide gauges.

A typical contour plot of significant wave height, Hs simulated using WAM over the Bay of Bengal at 0600 hours 29 December 2011 is projected in Fig. 10.6. The variation of H_s, T_m and θ_m at 20 m water depth is compared with that simulated using WAM during 14–31 December 2011 and is depicted in Fig. 10.7. It is inferred that H_s and θ_m are in good agreement under both normal and cyclonic duration near the study area. It is seen that the H_s varies from about 1.5 to 2.0 m, whereas, the T_m is found to be in the range of 6 s to 9 s until 96 hours before the landfall of cyclone. The said variations show a small degree of variability. Subsequently, a drastic increase of H_s and T_m demonstrates the severity of the sea state during the approach of Thane towards the coast. Its effect can be seen from the increase of H_s up to 6.2 m with an associated T_m up to 13 s in a water depth of 20 m. The predicted T_m deviates while the cyclone builds up the wave climate, i.e., just after 27 December. The details are reported by Anand et al. (2014).

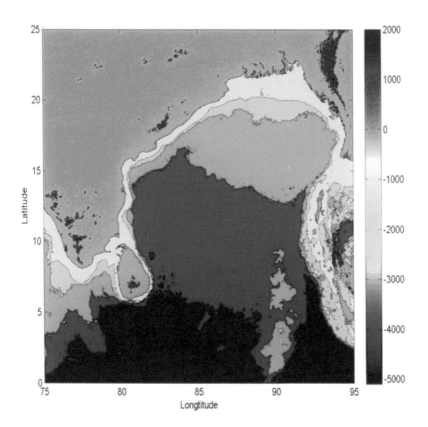

Fig. 10.4   Domain setup for WAM.

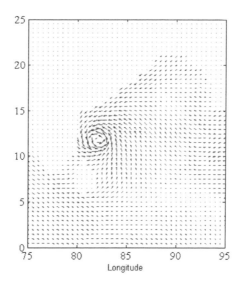

Fig. 10.5 Typical wind vector obtained from ENCEP at 0600 hrs, 29 December, 2011.

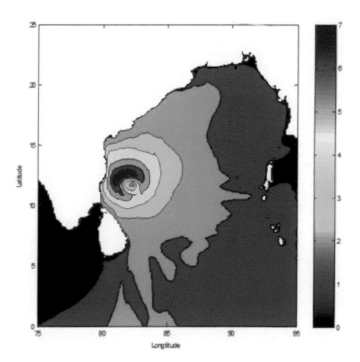

Fig. 10.6 Contour plot of significant wave height simulated using WAM at 0600 hours on 29 December 2011.

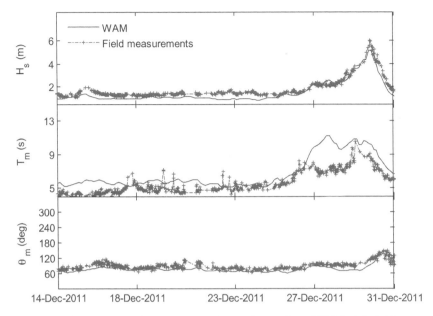

Fig. 10.7 Comparison of H_s, T_m and θ_m predicted from WAM with field measurements.

10.6 Mild-Slope Equation (MSE) Wave Models

In the shallow water environment, it is inevitable to take into account the variation in sea bottom topography for the wave propagation to accommodate the wave refraction. In these situations, to simplify the domain, vertically integrated equations derived from Euler equations are widely adopted based on the assumption of mildly varying depth of the seabed. The assumptions which for the derivation of mild-slope equation wave model are that of the linear waves and, the water depth is presumed to be gradually varying. The MSE has three different formulations: namely, the hyperbolic MSE for a time-dependent wave field; the elliptic MSE; and, the parabolic MSE for a steady-state wave field.

The elliptic model, put forth by Berkhoff in 1972, describes wave refraction and diffraction, both in deep and shallow waters. Though the MSE wave model is principally used to study monochromatic waves, it may be adapted to representing irregular waves by the summation of different wave harmonics. Mild-slope equation wave models are generally employed in the region ranging from a certain deep water location to the nearshore region, some distance away from the shoreline where non-linearity of waves is weak. The depth averaged mild-slope equation was derived by integrating the Laplace's equation over the depth, following the approach of Smith and Sprinks (1975). Even though MSE wave models have many practical advantages, this combined refraction-diffraction model demands immense computational efforts while dealing with short wave propagation in the nearshore region. This is due to that the nature of the equation, i.e., an inseparable elliptic partial differential equation. Application of this model in harbor resonance modelling is an exemption as the water depth, in this case would be high, rather than a mild sloped bed.

The combined refraction-diffraction equation that describes the propagation of periodic, small amplitude waves over an arbitrarily varying mild sloped sea beds Berkhoff (1972), is,

$$\frac{\partial}{\partial x}\left(CC_g \frac{\partial \phi}{\partial x}\right) + \frac{\partial}{\partial y}\left(CC_g \frac{\partial \phi}{\partial y}\right) + \sigma^2 \left(\frac{C_g}{C}\right)\phi = 0$$

$$\nabla(CC_g \nabla \phi) + \sigma^2 \left(\frac{C_g}{C}\right)\phi = 0 \qquad (10.4)$$

where, x and y are the two orthogonal co-ordinate directions in Cartesian system; $\phi(x,y,z) = \frac{\cosh k(t+d)}{\cosh kd}\phi(x,y)$; $\phi(x,y)$ is the complex velocity potential; σ = Angular wave frequency = $2\pi/T$, $C(x,y)$ = wave celerity = σ/k; $C_g(x,y)$ = group celerity; and $k(x,y)$ = wave number (= $2\pi/L$), related to the still water depth $d(x,y)$ through the dispersion relation, $\sigma^2 = gk\tanh(kd)$.

It is valid when the sea bed has a mild slope characterized by $\nabla d/kd = O(\varepsilon) \ll 1$ and that $O(\varepsilon^2)$ is neglected. The free surface elevation is described as below,

$$\eta(x,y) = -\frac{i\omega}{g}\phi(x,y)$$

Because of the elliptical nature of the above partial differential equation, a set of conditions at the whole boundary of solution must be given. By the substitution of the expression $[\phi = ae^{is}]$ and neglecting the variation of the amplitude function in the horizontal plane, the above equation gives rise to the refraction equation. In the case of deep waters or of constant depth, the above equation becomes Helmholtz diffraction equation. Equation (10.4) is transformed to the Helmholtz equation,

$$\nabla^2 \Phi + K^2(x,y)\Phi = 0 \qquad (10.5)$$

through the modified wave potential function, Φ and a modified wave number, K, as defined below.

$$\Phi = \phi(CC_g)^{0.5} \quad \text{and} \quad K^2 = k^2 - \frac{\nabla^2(CC_g)^{0.5}}{(CC_g)^{0.5}}$$

In the case of shallow water case, Eq. (10.5) reduces to long wave equation.

10.6.1 Solution of MSE

Let us consider a study domain bounded by Γ_2 with a large obstruction defined by Γ_1 as shown in Fig. 10.8. The boundary of the domain is assumed to be as a part of the circle to computationally ease the imposition of the far field boundary condition.

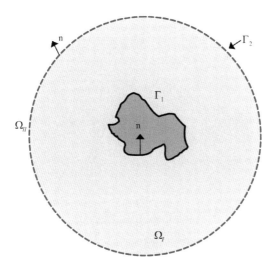

Fig. 10.8 Study domain for large body diffraction under long wave.

Now, it has to satisfy essentially two boundary conditions.

(a) Full/partial reflection from the obstruction bounded by Γ_1
(b) Radiation condition of infinity

Consider the domain into two distinct area with the demarcation indicative the inner zone Ω_I, that include the varying bottom contour within Γ_2 and, the outer zone, Ω_{II} is either of constant depth or deep water. Now, the solution in zone, Ω_{II} is the summation of the incident wave and the outgoing wave due to diffraction and refraction inside Ω_I. The outgoing wave leaving Γ_2 is considered as the wave emanating from the source, $q(S)$ at Γ_2 that satisfies the radiation boundary condition at infinity. The wave height and phase should be continuous across Γ_2. This condition leads to additional equations to determine the intensity of source distribution, $q(S)$.

The boundary conditions to solve within the domain, Ω_I are,

$$\Omega_I \begin{cases} \dfrac{\partial \varphi}{\partial n} = 0 \text{ at } \Gamma_1 \\ \dfrac{\partial \varphi}{\partial n} = f(\text{an orbitary function}) \text{ of } \Gamma_2 \end{cases} \quad (10.6)$$

In Ω_{II}, the solution must satisfy the Helmholtz equation of the diffraction problem and can be written in the form (i.e., solution at P in Ω_{II}) (Zienkiewicz et al., 1978).

$$\psi(P) = \tilde{\phi}(P) + \int_{\Gamma_2} q(S) \frac{1}{2i} H_0^1(kr) d\Gamma \quad (10.7)$$

where, $\tilde{\phi}(P)$ is due to incident wave; H_0^1 is the Hankel function of the first kind and zeroth order satisfying Helmholtz equation and the Sommerfeld condition at the farfield; and r is the distance between P and the point of intersect along Γ_2.

The source intensity function, q along Γ_2 must satisfy the integral equation,

$$\left(\frac{\partial \psi}{\partial n}\right)_P = \left(\frac{\partial \tilde{\phi}}{\partial n}\right)_P + q(P) + \int_{\Gamma_2} q(S) \frac{\partial}{\partial n} \left[\frac{1}{2i} H_0^1(kr)\right] d\Gamma \quad (10.8)$$

While $r = 0$, i.e., P is on Γ_2, the following continuity conditions implies,

$$\Psi = \phi$$

and, $\frac{\partial \Psi}{\partial n} = \frac{\partial \varphi}{\partial n}(=f)$. Now on Γ_2, the problem is well defined and the unknown functions q and φ can be evaluated. Once the intensity of source distribution, q known, solution $\psi_{\Omega_{II}}$ can be obtained at all "P".

Most of the currently available public domain and commercial codes numerically solve the mild slope equation using grid based finite difference or finite element methods. The following section briefly explains the solution procedure using finite difference method.

Finite difference method

Let us consider a rectangular domain, ABCD in which the waves enter the domain from the left boundary, AC and the right boundary, BD falls on the shoreline. Finite difference operator replaces the differential operator in Eq. (10.8).

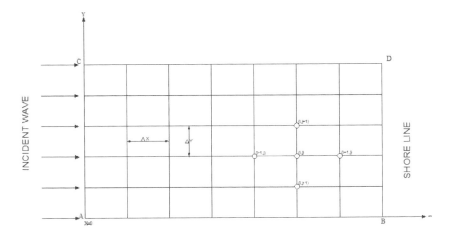

Fig. 10.9 Typical finite difference grid.

The boundary conditions to be imposed along AB, AC, BD, CD and in general, they may be written as,

$$L(\Phi) = 0$$

The total potential can be expressed as $\Phi = \Phi_{in} + \Phi_r$. Along AC, the incident wave is represented by $[\Phi_{in} = A_{in} \exp(iKx)]$, $i = \sqrt{-1}$. There exists also a backscattered component, which may be approximated by $[\Phi_r = B \exp(-iKx)]$ [Booij (1983)]. "B" is not necessarily known, the imposition of flow boundary conditions leads to the following equations.

$$\frac{\partial \Phi}{\partial x} = \frac{\partial \Phi_{in}}{\partial x} + \frac{\partial \Phi_r}{\partial x}$$

$$\frac{\partial \Phi}{\partial x} = \frac{\partial \Phi_{in}}{\partial x} - iK\Phi_r$$

$$\frac{\partial \Phi}{\partial x} = iK \cos(\theta)\Phi_{in} - iK(\Phi - \Phi_{in})$$

$$\frac{\partial \Phi}{\partial x} = iK[(1 + \cos\theta)\Phi_{in} - \Phi]$$

where, θ is the wave angle. The other boundaries may represent a breakwater (an obstruction), an open boundary, or a coastline. In such cases the following condition is imposed [Tsay and Liu, 1983].

$$\frac{\partial \Phi}{\partial n} - iK\alpha\Phi = 0 \qquad (10.9)$$

where n is the direction normal to the boundary and α is a relaxation coefficient that varies with the type of boundary and may have to be determined empirically. The values of "α" varies from 0 to 1 ($\alpha = 0$ for full reflection and $\alpha = 1$ for full absorption). The incident wave angle, θ in the offshore open boundary will be obtained as the transformation from offshore wave climate to shallow water wave angles by applying the Snell's law. The domain is discretized into grids of size (Δx) and (Δy). A typical finite difference grid is shown in Fig. 10.9. $\{\Phi_{(x,y)}\}$ is used to donate the grid point value of the potential, standard discretization of the Helmholtz equation using second order finite difference scheme yields,

$$\frac{\Phi_{(i-1,j)} + \Phi_{(i+1,j)}}{\Delta x^2} + \frac{\Phi_{(i,j-1)} + \Phi_{(i,j+1)}}{\Delta y^2}$$

$$- \left[\frac{2}{\Delta x^2} + \frac{2}{\Delta y^2} - K^2 \right] \Phi_{(i,j)} = 0 \qquad (10.10)$$

The conventional approach consists of writing such equations for all points in the domain. The resulting system of equations may be expressed in matrix form as:

$$[A]\{\Phi\} = \{f\} \qquad (10.11)$$

where $[A]$ is the system matrix, $\{\Phi\}$ is the unknown vector (of the desired grid point value of the wave potential), and $\{f\}$ is a vector that contains information about the discretized boundary condition. The direct methods like Gauss elimination requires large amount of memory to store the system matrix values which makes it practically impossible, whereas the iterative methods do not require the storage of matrix $[A]$ and hence can be used for large domains. In order to accomplish this, the matrix $[A]$ must be strictly diagonally dominant, or it must be symmetric and positive definite. Conventional iterative methods like Jacobi's or Gauss-Siedel methods, do not guarantee convergence when applied to Eq. (10.11) since the matrix obtained is neither diagonally dominant nor positive definitive (because of complex quantities in the boundary conditions). An example for the iterative scheme is the Generalized Conjugate Gradient method, which has proven to converge several orders of magnitude faster than any direct schemes. The algorithm is as given below.

1. Select trial values Φ_0 (i.e. $i = 0^{\text{th}}$ iteration) for all grid points where the solution is desired.
2. Compute for all points: $[p_0 = r_0 = f - A\Phi_0]$, $[u_i = Ap_i]$
3. Compute for the ith iteration: $\alpha_i = \dfrac{r_i^T A r_i}{u_i^T u_i}$
4. Update $\Phi_{i+1} = \Phi_i + \alpha_i p_i$, for all points
5. Check for convergence of solution
6. Compute for each grid point: $[r_{i+1} = r_i - \alpha_i A p_i]$, $[r_{i+1} = r_i - \alpha_i u_i]$
7. Compute for i^{th} iteration: $\beta_i = \dfrac{r_{i+1}^T A r_{i+1}}{r_i^T A r_i}$
8. Compute: $[p_{i+1} = r_{i+1} + \beta_i p_i]$
9. Set $i = i + 1$, and go to step 3.
10. The procedure is convenient even for non-rectangular domains, since the algorithm simply hops from one grid point to the next.

Convergence criteria

The convergence criteria used here is: $\dfrac{\sum |(\nabla^2 \Phi + K^2 \Phi)|^2}{\sum |\Phi^2|} < \varepsilon$

Where the summations extend over all the grid-points and ε is the prescribed tolerance limit.

10.7 Boussinesq Approximation

The Boussinesq approximation is valid for long waves. This approximation satisfies only weakly nonlinearity. Essentially it is obtained by approximating the vertical structure of the flow velocity. The resulting non-linear partial differential equations are called as the Boussinesq equations. These equations incorporate frequency dispersion unlike in shallow water equations in which, the speed of the wave depends on the bottom topography irrespective of frequency of the wave. Due to this, Boussinesq equations can better model the nearshore waves and also, wave penetration into harbours can be modelled by considering the effects of diffraction, bottom refraction and shoaling.

The approximation of vertical structure of flow under water waves has been achieved since the waves propagate in the horizontal direction with the harmonic variation in both horizontal directions. However, the variation across the depth has different fitting. During the above process, the modelling of vertical coordinate is avoided. Thus, the three-dimensional fluid domain is simplified into two-dimensional horizontal domain. This reduces one order of the number of equations to be solved.

The following procedure explains the Boussinesq approximation mathematically.

1. At an elevation, a Taylor series expansion is made on the velocity potential and/or horizontal and vertical water particle velocities.
2. Similar to the solution procedure of any infinite series, only a finite number of terms is selected without omitting any terms from the first term. That is, n^{th} term was not selected without selecting $(n-1)^{\text{th}}$ term.
3. In the Taylor expansion, the partial derivatives corresponding to vertical coordinates are replaced with partial derivatives with respect to horizontal derivatives. While doing so, the conservation of mass following incompressible fluid assumption is ensured and the irrotationality of flow is ensured by enforcing zero curl condition.
4. Final partial differential equations are represented only in terms of the horizontal coordinates and time for dynamic problems.

10.7.1 Boussinesq equations

Consider the potential flow problem in 2D (x-horizontal and z-vertical with the origin at left and at still water level). The water depth (d) is constant. Write the Taylor series expansion for ϕ at $z = -d$ (let be, ϕ_b).

$$\phi = \phi_b + z \left[\frac{\partial \phi}{\partial z}\right]_{z=-d} + \frac{1}{2} z^2 \left[\frac{\partial^2 \phi}{\partial z^2}\right]_{z=-d}$$

$$+ \frac{1}{6} z^3 \left[\frac{\partial^3 \phi}{\partial z^3}\right]_{z=-d} + \frac{1}{24} z^4 \left[\frac{\partial^4 \phi}{\partial z^4}\right]_{z=-d} \quad (10.12)$$

For the incompressible flow, $\frac{\partial \varphi}{\partial d} = 0$ at the impermeable bed. Hence,

$$\phi = \left\{\phi_b - \frac{1}{2} z^2 \frac{\partial^2 \phi_b}{\partial x^2} + \frac{1}{24} z^4 \frac{\partial^4 \phi_b}{\partial x^4} + \cdots\right\} + \left\{z \left[\frac{\partial \phi}{\partial z}\right]_{z=-d} - \frac{1}{6} z^3\right\}$$

$$= \left\{\phi_b - \frac{1}{2} z^2 \frac{\partial^2 \phi_b}{\partial x^2} + \frac{1}{24} z^4 \frac{\partial^4 \phi_b}{\partial x^4} + \cdots\right\} \quad (10.13)$$

The above series can be truncated according to the required accuracy.

Now, apply the kinematic and dynamic free surface conditions in fully nonlinear form.

$$\frac{\partial \eta}{\partial t} + u \frac{\partial \eta}{\partial x} - w = 0$$

$$\frac{\partial \phi}{\partial t} + \frac{1}{2}(u^2 + w^2) + g\eta = 0 \quad (10.14)$$

Limited up to quadratic terms with respect to η and the velocity potential expanded at the given elevation, ϕ_b, the following equations can be derived. It is here assumed that cubic and higher order terms are negligible and hence, posing weak nonlinearity.

$$\frac{\partial \eta}{\partial t} + \frac{\partial}{\partial x}[(d+\eta)u_b] = \frac{1}{6}d^3 \frac{\partial^3 u_b}{\partial x^3},$$

$$\frac{\partial u_b}{\partial t} + u_b \frac{\partial u_b}{\partial x} + g\frac{\partial \eta}{\partial x} = \frac{1}{2}d^2 \frac{\partial^3 u_b}{\partial t \partial^2 x} \quad (10.15)$$

In the above equations, if the right-hand side terms are set to be zero, then, the resulting equations are called as "shallow water equations".

Combining the above two equations with a linear approximation of u_b (horizontal flow velocity at $z = -d$) results the following:

$$\frac{\partial^2 \psi}{\partial T^2} - \frac{\partial^2 \psi}{\partial \xi^2} - \frac{\partial^2}{\partial \xi^2}\left(\frac{1}{2}\psi^2 + \frac{\partial^2 \psi}{\partial \xi^2}\right) = 0 \quad (10.16)$$

The linear frequency dispersion characteristics of Boussinesq equations without linear approximation of bottom particle velocity is,

$$C^2 = gd\frac{1 + \frac{k^2 d^2}{6}}{1 + \frac{k^2 d^2}{2}} \quad (10.17)$$

where, C is the phase speed and k is the wave number.

For an approximated Boussinesq equation [Eq. (10.18)], the linear frequency dispersion equation can be derived as,

$$C^2 = gd\left(1 - \frac{k^2 d^2}{3}\right) \quad (10.18)$$

Considering a relative error of 4% in the phase speed compared with linear wave theory estimate, Eq. (10.17) is valid for $kd < \pi/2$ and Eq. (10.18) is valid for $kd < 2\pi/7$ for engineering applications. That is the former equation is valid for wavelengths larger than four times the water depth and the approximated form is valid for wavelengths larger than seven times the water depth. That is, the latter is valid for very long waves. Depending on the applications, the simplified form of Boussinesq equation can be adopted. However, the shallow water equations provide an estimate of wave length with less than 4% accuracy is for the condition of wavelengths larger than 13 times the water depth. The above clearly states the regime of applications of various forms of Boussinesq approximation.

Any form of modification to Boussinesq equations, either in the form of domain, i.e., varying bathymetry or higher order accuracy terms in the series expansion or the incorporation of additional physics such as wave breaking, surface tension, nonlinear interaction are generally represented as Boussinesq-type equations. However, in engineering applications, the equations are often extended beyond the breaking point, up to wave run-up in the swash zone. This is possible by adding an artificial energy dissipation term for wave breaking. Besides, efforts are also made to extend the model to deeper water. Unlike the wave spectral model and the MSE model, the Boussinesq model does not have the presumption that the flow is periodic. Therefore, it can be applied to waves induced by impulsive motions, i.e., solitary waves, landslide-induced waves, tsunami, and unsteady undulation in open channels (Lin, 2008).

Some of the simplified form of Boussinesq equations based on the assumption of waves travel along one direction have specific applications as listed below.

a. Equation of wave propagation in one dimension is called Korteweg-de Vries equation: both non-periodic solitary waves and periodic cnoidal waves can be derived from kdV equation, i.e., approximated solutions of the Boussinesq equations.
b. Equation of wave propagation in two dimension is called Kadomtsev-Petviashvili equation
c. The nonlinear Schrödinger equation (NLS equation) is for the complex valued amplitude of narrowband waves.

10.7.2 Shallow-water equation wave models

To model tsunami or other long waves (e.g., tides), a shallow-water equation (SWE) model is more likely to be adopted. Compared with the Boussinesq model, the SWE model is simpler because the flow is assumed to be uniform across the water depth and the wave-dispersive effect is neglected. The SWE model has a wide application range in modelling tsunami, tides, storm surges, and river flows. The main limitation of the SWE model is that it is suitable only for flows whose horizontal scale is much larger than vertical scale (Lin, 2008).

References

Anand K. V., Sannasiraj, S. A. and Sundar, V. (2014). Investigation on the cyclonic sea state along Southeast Coast of India, *Marine Geodesy*, 38, 58–78.
Berkhoff, J. C. W. (1972). Computation of combined refraction-diffraction, *Proc. 13th Int. Conf. Coastal Engineering*, pp. 471–90.
Booij, N. (1983). A note on the accuracy of the mild slope equation, *Coastal Eng.*, 7, 191–203.

Booij, N., Ris, R. C. and Holthuijsen, L. H. (1999). A third-generation wave model for coastal regions: 1. Model description and validation, *J. Geophys. Res.*
Bretschneider, C. L. (1952). The generation and decay of wind waves in deep water, *Trans. A.G.U.*, 33(3), 381–389.
Bretschneider, C. L. (1958). Revision in wave forecasting: Deep and shallow water, *Proc. 6th Conf. on Coastal Eng.*, 30–67.
Hasselmann, S., et al. (1988). WAMDI Group, The WAM model — A third generation ocean wave prediction model, *J. Phys. Oceanogr.*, 1.i, 1775–1810.
Lin, Pengzhi (2008). *Numerical Wave Modelling*, Taylor and Francis, London.
Ris, R. C., Holthuijsen, L. H. and Booij, N. (1999). A third-generation wave model for coastal regions: 2. Verification, *J. Geophys. Res.*, 104, 7667–7681.
Shore Protection Manual, Volume 1. (1977). *Beach Erosion.* Department of the Army, Coastal Engineering Research Center.
Smith, R. and Sprinks, T. (1975). Scattering of waves by a conical island, *J. Fluid Mech.*, 72, 373–84.
Sverdrup, H. U. and Munk, W. H. (1947). Wind, sea and swell. Theory of relations for forecasting, U.S. Navy Hydrographic Office, Washington, Pub. No. 601, 44 pp.
Tsay, T.-K. and Liu, P.L.-F. (1983). A finite element model for wave refraction and diffraction, *Applied Ocean Research*, 5(1), 30–37.
Zienkiewicz, O. C., Lewis, R. W. and Stagg, K. G. (eds.) (1978). *Numerical Methods in Offshore Engineering*, Wiley.

Index

A

accretion, 12, 13, 139
armour units
    Accropode, 170, 171, 188, 242
    antifer cube, 171
    CORE-LOC, 188, 242
    cob, 171
    concrete, 170
    Dolos, 170–172, 175, 176, 178–181, 188, 236
    hollow cube, 188
    Rock, 143, 188
    Tetrapod, 169–172, 188, 236, 242, 253, 257, 278, 279
Airy's wave theory, 31, see wave; Airy
alongshore
    current velocity, 75
    sediment transport, 64
alongshore drift, 133
angle of repose, 28
angular wave frequency, 328
approach channel, 26, 110, 116, 148, 150–152, 303
armour layer, 160, 165, 180, 183, 253
artificial beach nourishment, 137
artificial reefs, 103, 113
average overtopping rate, 202

B

backscattered component, 331
backshore, 1, 13, 106, 114, 115
barrier
    beaches, 17
    islands, 15, 17
base discharge, 200

Bathymetry, 2, 6, 9, 38, 68, 70, 102, 165, 323
bay, 7, 13, 15, 68, 143, 144
beach, 1, 42, 115
    nourishment, 114, 116, 117
    profile, 1, 2, 41, 119, 130
    ridges, 17
    sands, 12, 14, 31, 58
    slope, 8, 50, 53, 301
    width, 56
bed
    load, 31, 38–40, 51, 55, 108, 215
    sediment, 24, 25, 31, 216
    shear stress, 39, 230–232
berm, 42, 189, 160, 190, 196
    reduction coefficient, 203
    sand, 2
    width, 202
bottom particle velocity, 335
boundary layer, 214
Boussinesq
    approximation, 333, 335
    equation, 334, 335
breaker
    line, 42, 57, 58, 61, 62, 102
    parameter, 190, 202
    zone, 1, 8, 42, 46, 47, 49, 55
breaking point, 51, 187, 336
breakwater, 6, 18, 97, 99, 103, 159, 160
    berm, 166
    caisson, 161, 167, 279
    composite, 161, 164, 237, 246, 256
    detached offshore, 159

horizontally composite, 257
mound, 160
semi-circular, 162
semi-circular detached, 167
sloping, 160
submerged, 13, 101, 159
submerged offshore, 105
S-type, 167
vertical, 160–162
Buckingham's pi theorem, 299, 301
buried toe, 229
bypassing, 144, 145, 148, 150

C

caisson, 106, 162, 246, 256
cell circulation, 53
channel, 21, 26, 31, 68, 137, 148, 218
clapotis, 237, 243, 245
cliff, 15, 42, 235
coastal
    defence, 189, 197
    environment, 1, 67
    erosion, 10, 97, 98, 122
    features, 15, 17
    management, 10
    morphology, 1, 64, 66, 102
    protection, 1, 10, 99, 101, 102, 159, 235
    structures, 69, 185, 197, 235, 302
coastline, 9, 47, 64, 68, 117, 121, 332
coefficient
    berm reduction, 203
    curvature, 22
    uniformity, 22
    drag, 26, 118, 232
    friction, 28, 241
    layer, 275, 276, 278
    reflection, 237, 242
    refraction, 6
    roughness, 188–189, 195
    shoaling, 207
    stability, 169, 172, 180, 182
cofferdams, 217
complex slopes, 186, 202

composite slopes, 189
composite structures, 202
convergence criteria, 333
coral reefs, 113, 124, 125
critical shear stress, 33, 34, 39, 231
crown wall, 106, 136, 160, 165
currents, 2, 38, 41, 50, 55, 64, 101, 118, 145, 185, 226, 232, 305
current drag, 264
current friction factor, 62, 82

D

damage assessment, 182, 183
    area method, 182
    number of units method, 183
dampening, 18
Darcy' law, 301
deformations, 2
depth limited, 7
domain
    frequency, 320
    time, 189
diffraction, 6, 7, 9, 105
dimensionless stability, 242
discharge, 18, 38, 145, 146, 194, 196–198
    factor, 197–200
    rate, 144
    velocity, 154
dispersion relation, 328
downdrift, 110, 118, 135, 148–150, 152, 229
drag, 21, 26, 29, 229
    coeffcient, 26, 118, 232
    force, 39, 299, 300
drift, 55, 121, 134, 137, 144, 148, 150, 153, 154
dune, 14, 98
    instability, 67
dynamic
    forces, 236
    oceanic system, 20
    pressure, 45, 236, 246
    similarity, 305–307
    similitude, 304

E

earthquakes, 21, 30, 32, 265
eddy, 148, 224
effective
  depth, 264
  stress, 32, 33
  width, 203
electromagnetic forces, 21
electrostatic forces, 215
energy dissipation, 52
energy flux, 3, 56, 58, 135
  method, 122
  parameter, 60
entrainment velocity, 31
environmental
  forces, 212
  impact, 143
equilibrium velocity, 26
equivalent
  diameter, 275
  length, 264
  slope, 197, 200
erosion, 100, 115
estuaries, 15, 17, 38, 68, 116, 144, 303
estuarine conditions, 119
extreme events, 21, 67, 97, 115, 119, 159

F

failure mode, 132, 175, 229, 257
fall velocity, 26, 27, 29, 156
fetch, 13
field investigations, 11
filter layer, 132, 165, 236, 254, 255, 276, 279
filter mats, 112
fishtail, 108
flooding, 68, 130, 185, 194, 204, 235
flood
  currents, 145
  inundation, 186
flow velocity, 212
fluctuations, 44
fluid mechanism of scour, 212

force, 28
  equilibrium, 48
  reduction factor, 243
foreshore, 1, 55, 133
freeboard, 195, 198, 280
frequency
  dispersion, 333, 335
  domain, 320
friction, 2, 20, 22, 31, 113, 165, 259
  coefficient, 28, 241
  factor, 61, 227, 228, 242
  forces, 256
  stress, 50
Froude number, 306

G

gabion, 106, 139, 143, 259
  boxes, 130, 132
  mattress, 106
generalized conjugate gradient
  method, 332
geosynthetics, 108, 110, 112, 130, 143, 229
  geobags, 139, 143
  geocomposites, 112
  geocontainers, 143
  geogrids, 112, 260
  geomembranes, 112
  geotextiles, 112, 139, 229
  geotube, 106, 139, 143, 159
Goda's method of breaking wave
  force, 251
grid
  node, 319
  points, 53, 318, 319, 332, 333
groin, 6, 10, 99, 106, 122, 185
  construction, 108
  fields, 11, 134, 108, 109, 119, 121

H

Hankel function, 330
harbour, 12, 24, 25, 118, 121, 137, 152, 159, 164, 165, 295
  breakwaters, 118
  walls, 172

hard measures, 102, 119, 143
hazard, 67
headlands, 13, 65–67, 99, 144, 160
Helmholtz equation, 328, 330
high-head scour, 218
horizontal force, 39, 265
horseshoe vortex, 212–214
Hudson's formula, 169, 181
hurricane, 71, 74
hydraulic drag, 170
hydrostatic, 45, 247, 252
    loading, 237
    moment, 247
    pressure, 45, 238, 247

I

immersed weight transport rate, 58
impermeable structures, 191, 198
infrastructure, 68
inlet, 145
inshore, 51, 116
integrated
    pressure, 44
    velocity fluctuations, 44
    wave characteristics, 70
Iribarren number, 187

J

jetty, 110, 148, 169, 264

K

Kamphuis method, 93
Kolos, 172, 175, 176, 179, 181

L

lagoons, 15, 17
Laplace's equation, 328
layer
    coefficient, 275, 276, 278
    thickness, 185, 190
lee side, 167, 185, 214, 253, 258
LEO – littoral environment observation, 60
lift, 21, 39
linear wave theory, 2

liquefaction, 30, 32, 33
littoral
    barriers, 118
    drift, 11, 12, 54, 102, 103, 117, 145, 148, 150
    transport, 11, 40, 41
longshore
    current, 53, 55, 62, 63, 106, 135
    drift, 110, 135, 150
    energy flux factor, 87
    energy flux, 58, 60
    sediment transport, 11, 13, 53, 57, 78, 106
    transport, 11, 56, 61
long waves, 9, 333, 336
low tide level, 118

M

maintenance dredging, 26, 137
marine piles, 264
mean
    discharge, 195, 198
    overtopping discharge, 200, 201, 235
    sea level, 13, 185
median
    diameter, 232
    grain size, 23, 52
    particle size, 23
mild slope equation, 327, 328, 330
Minikin method, 243, 245
moisture content, 25
momentum flux, 40, 50, 191
morphological changes, 10
MSL, 15, 160, 187, 197

N

navigation, 145, 147, 218
navigational channel, 20, 148, 150, 159, 235
net
    drift, 40, 119
    sediment drift, 122
nonlinear Schrödinger equation, 336

non-breaking waves, 187, 220
nourishment, 115

O

offshore
  breakwaters, 103, 104
  detached breakwaters, 11
  reefs, 13
orbital velocity, 41, 221
oscillating water column, 161
overtopping, 105, 132, 185, 194, 196, 204, 242
  discharge, 194, 195, 198

P

partial safety factors, 235
particle
  acceleration, 8, 297
  orbit, 297
  size correction factor, 64
  size, 21, 28
  velocity, 8, 297
peak
  discharges, 204
  frequency, 322
perennial erosion, 97, 119, 120, 130
permeability, 29, 30
phase
  speed,335
  averaged, 320
  resolving, 320
physical model, 104, 159, 160, 172, 195, 303, 307, 309
piles, 210, 219
  marine, 264
  raking, 264
  sheet, 230, 260–261
  steel, 264
pipelines, 117, 148, 222, 224, 227
plain cube, 171
plasticity, 25
plasticity index, 25, 112
plunger breakers, 8, 62, 253
porosity, 29, 185, 254
P-M spectrum, 309

R

radiation
  boundary condition, 330
  condition, 329
  stresses, 43, 45, 48
rainfall, 10, 15
Rayleigh distribution, 57, 58
recirculating eddy, 213
reduction factor, 196
reefs, 13, 99, 127, 143
  artificial, 103, 113
  coral, 113, 124–125
  offshore, 13
  rock, 144
  submerged, 147
Reynolds
  number, 26, 33, 227, 300, 306
  stresses, 50
roughness
  coefficient, 188–189, 195
  length, 231, 232
  reduction factor, 203
rubble mound, 106, 143, 172
  breakwaters, 162, 165, 167, 169
  groins, 122
  structures, 97, 159, 241, 252
run-up, 52, 105, 122, 130, 186, 187, 189, 191, 193, 206

S

safety factors, 257, 258
salient, 103–105, 126
sand, 14
  bar, 111, 145, 148, 153
  bypassing, 147
  dunes, 68
  traps, 147
scale effect, 302, 305
scour, 210
  boat, 218
  degradation, 217
  failure, 219
  general, 216
  global or dishpan, 218

high-head, 218
hole, 220, 222, 224, 227
local, 216, 217, 219
maximum depth, 220–221, 227–228
mechanism, 211
protection, 228–229
sea
  state, 71, 241, 251, 295, 325
  level
    mean, 13, 185
    rise, 67, 99
    change, 15
sediment
  bed, 24–25, 31, 216
  cohesionless, 215
  cohesive, 215
  net drift, 122
  budget, 218
  cell, 64–67
  concentration, 28, 41
  dynamics, 69, 112, 215, 295
sediment transport
  cross-shore, 50, 57, 186
  longshore/alongshore, 11, 13, 53, 57, 78, 106
  volume, 58
shape factor, 30
shields parameter, 34–35, 37, 215, 231
shore protection manual, 11, 42, 142
sills, 217, 230
similitude, 303
  Geometric, 303
  Dynamic, 304
  Kinematic, 304
SMB prediction curves, 322
Snell's law, 5, 332
spectrum
  directional spectra, 70, 324
  P-M, 309
  shape, 185
  wave, 321
spits, 15, 17, 65
spurs, 148, 159
stability
  coefficient, 169, 172, 180, 182
  number, 181, 253
  parameters, 180
stagnation
  line, 226
  pressure, 213
storm, 50, 67, 98, 101
  cyclone, 10, 69
  size, 68
  surge, 21, 50, 67–68, 144
  tide, 68
  water, 68
  wave, 30
surf
  width, 75, 76, 119, 148
  zone, 40–42, 51–52, 54, 56, 62
  zone width, 63, 107–108, 119
surface roughness, 169, 190
suspended
  load, 31, 38, 40
  particles, 20
swan, 321
swash, 53, 187
  zone, 51–52, 336
swell
  correction, 64
  waves, 13

T

tidal
  discharge, 145
  flushing, 111, 144
  inlets, 111, 143, 145, 147, 150
  prism, 145
  range, 38, 104, 119, 159
tide, 43
  gauge, 69
  level, 42
tombolo, 103–104
training walls, 111, 147, 149–150, 154, 235
tranquil conditions, 253
tsunami, 12, 97, 100, 118, 122, 336
types of breakers
  collapsing breakers, 8
  plunging breakers, 8, 253

spilling breakers, 8
surging breakers, 8, 254

U

updrift, 110, 116, 118, 148, 150
uplift, 229–230, 251

V

velocity
    alongshore current, 75
    bottom particle, 335
    discharge, 154
    entrainment, 31
    equilibrium, 26
    fall, 26–27, 29, 156
    flow, 212
    longshore current, 62–63, 135
    orbital, 41, 221
    particle, 8, 297
    potential, 334–335
    water particle, 4

W

WAM, 321, 324
water particle
    acceleration, 4
    displacement, 4
    velocity, 4, 334
wave, 70, 97, 106, 172, 242
    Airy, 31
    attack factor, 203
    breaking, 1, 63, 106, 256, 321, 336
    diffraction, 6, 7, 9, 105
    direction, 3, 5, 6, 57, 122, 160, 185, 280
    energy convertor, 161
    energy, 46, 56, 75, 103, 105, 113, 119, 126, 160, 165, 170, 190, 253, 257, 321

    friction factor, 231
    momentum flux, 192
    prediction, 323
    reflection, 8, 9, 161, 332
    refraction, 5–6, 47, 61, 97
    rider buoy, 69
    set-up, 40, 187
    shoaling, 3, 61, 150, 333
    spectrum, 321
    steepness, 8, 50, 61, 190
    -wave interaction, 321
    types/classification
        breaking, 63, 106, 256, 321, 336
        gravity, 126, 304
        long, 9, 333, 336
        long-crested, 203
        non-breaking, 187, 220
        short-crested, 203
        solitary, 336
        standing, 8, 237
        swell, 13
wedge, 263
weirs, 148, 217, 306
wind
    direction, 322
    field, 322
    forces, 43, 264
    intensity, 68
    profile, 68
    radii, 68
    speed, 69, 74, 78
    vector, 324

Z

zero moment wave height, 190
zero shear point, 286